T0212259

Reconfigurable Antennas

Reconfigurable Antennas
Jennifer T. Bernhard

ISBN: 978-3-031-00407-0   paperback
ISBN: 978-3-031-01535-9   ebook

DOI: 10.1007/978-3-031-01535-9

A Publication in the Springer series

SYNTHESIS LECTURES ON ANTENNAS #4

Lecture #4

Series Editor: Constantine A. Balanis, Arizona State University

Series ISSN

ISSN 1932-6076 print
ISSN 1932-6084 electronic

Reconfigurable Antennas

Jennifer T. Bernhard
University of Illinois at Urbana–Champaign

SYNTHESIS LECTURES ON ANTENNAS #4

To Bill, Raine, and Ezra

ABSTRACT

This lecture explores the emerging area of reconfigurable antennas from basic concepts that provide insight into fundamental design approaches to advanced techniques and examples that offer important new capabilities for next-generation applications. Antennas are necessary and critical components of communication and radar systems, but sometimes their inability to adjust to new operating scenarios can limit system performance. Making antennas reconfigurable so that their behavior can adapt with changing system requirements or environmental conditions can ameliorate or eliminate these restrictions and provide additional levels of functionality for any system. For example, reconfigurable antennas on portable wireless devices can help to improve a noisy connection or redirect transmitted power to conserve battery life. In large phased arrays, reconfigurable antennas could be used to provide additional capabilities that may result in wider instantaneous frequency bandwidths, more extensive scan volumes, and radiation patterns with more desirable side lobe distributions. Written for individuals with a range of experience, from those with only limited prior knowledge of antennas to those working in the field today, this lecture provides both theoretical foundations and practical considerations for those who want to learn more about this exciting subject.

KEYWORDS

reconfigurable antennas, multifunction antennas, reconfigurable apertures, multifunction apertures

Contents

CHAPTER 1

Introduction

Antennas are necessary and critical components of communication and radar systems. Arguably, nine different types of antennas have proliferated during the past 50 years in both wireless communication and radar systems. These nine varieties include dipoles/monopoles, loop antennas, slot/horn antennas, reflector antennas, microstrip antennas, log periodic antennas, helical antennas, dielectric/lens antennas, and frequency-independent antennas. Each category possesses inherent benefits and detriments that make them more or less suitable for particular applications. When faced with a new system design, engineers change and adapt these basic antennas, using theoretical knowledge and general design guidelines as starting points to develop new structures that often produce acceptable results.

Nevertheless, the choice of an antenna from the families mentioned above also imposes restrictions on the overall system performance that arises because the antenna characteristics are fixed. Making antennas reconfigurable so that their behavior can adapt with changing system requirements or environmental conditions can ameliorate or eliminate these restrictions and provide additional levels of functionality for any system.

1.1   WHAT CONSTITUTES RECONFIGURABILITY?

Reconfigurability, when used in the context of antennas, is the capacity to change an individual radiator's fundamental operating characteristics through electrical, mechanical, or other means. Thus, under this definition, the traditional phasing of signals between elements in an array to achieve beam forming and beam steering does not make the antenna "reconfigurable" because the antenna's basic operating characteristics remain unchanged in this case [1, 2].

Ideally, reconfigurable antennas should be able to alter their operating frequencies, impedance bandwidths, polarizations, and radiation patterns independently to accommodate changing operating requirements. However, the development of these antennas poses significant challenges to both antenna and system designers. These challenges lie not only in obtaining the desired levels of antenna functionality but also in integrating this functionality into complete systems to arrive at efficient and cost-effective solutions. As in many cases of technology development, most of the system cost will come not from the antenna but the surrounding technologies that enable reconfigurability.

1.2 WHY WOULD ANTENNA RECONFIGURABILITY BE USEFUL?

Let us examine two general application areas, single-element scenarios and array scenarios, to motivate our later discussions of antenna reconfigurability. Straightforward single-element scenarios include portable wireless devices, such as a cellular telephone, a personal digital assistant, or a laptop computer. Single antennas typically used in these devices are monopole or microstrip antenna based and may or may not have multiple-frequency capabilities. Some packages may use two or three antennas for diversity reception on small devices to increase the probability of receiving a usable signal, but usually only one of the antennas is used for transmission. The transmission from the portable device to a base station or other access point is the weakest part of the bidirectional communication link because of the power, size, and cost restrictions imposed by portability. Moreover, the portable device is often used in unpredictable and/or harsh electromagnetic conditions, resulting in antenna performance that is certainly less than optimal. Antenna reconfigurability in such a situation could provide numerous advantages. For instance, the ability to tune the antenna's operating frequency could be utilized to change operating bands, filter out interfering signals, or tune the antenna to account for a new environment. If the antenna's radiation pattern could be changed, it could be redirected toward the access point and use less power for transmission, resulting in a significant savings in battery power.

The second application area, antenna arrays, consists of complex structures with an extensive history and a well-established set of limitations. For instance, current planar phased array technology is typically limited in both scan angle and frequency bandwidth as a result of the limitations of the individual array elements and the effects of antenna element spacing. Many of these established applications assume that the antenna element pattern is fixed, all of the elements are identical, and the elements lie on a periodic grid. The addition of reconfigurability to antenna arrays can provide additional degrees of freedom that may result in wider instantaneous frequency bandwidths, more extensive scan volumes, and radiation patterns with more desirable side lobe distributions.

When one considers adding new features to existing antennas that challenge the norm, inevitably (and rightly), questions arise about the true benefits and costs of doing so. Therefore, it is useful to discuss briefly the factors that will determine whether antenna reconfigurability makes sense in each new application.

Certainly, addition of antenna reconfigurability creates the need for more complex control. Full exploitation of antenna functionality may require dedicated signal processing and feedback circuitry and will certainly entail inclusion of more components and more complicated fabrication procedures compared with a traditional antenna system embodiment. However, if a single reconfigurable antenna could deliver the same functionality of more than one traditional single-purpose system, significant savings in cost, weight, volume, and maintenance/repair resources may be realizable. Of course, integrating new kinds of functionality into antennas will not automatically result in higher or comparable performance and lower costs—system designers need to be willing to exploit these new degrees of freedom and functionality so that the antenna becomes a more active part of the communication link, working together with new circuits, signal processing techniques, and communication and radar protocols.

Chapter 2 describes in detail the critical parameters for antenna operation, and Chapter 3 describes the linkages between these parameters that must be addressed during the development of reconfigurable antennas. Chapters 4–6 expound on possible methods for achieving reconfigurability in frequency, polarization, and radiation patterns, respectively. Each chapter also contains recent examples of published work to both illustrate the theoretical concepts and provide insight into the potential and limitations of each approach. Chapter 7 describes efforts to achieve *compound reconfigurability*, the combination of reconfigurability in multiple operating dimensions. Chapter 8 discusses practical issues surrounding the implementation of reconfigurable antennas, including trade-offs in reconfiguration mechanisms and their selection, as well as bias and control line design. Finally, Chapter 9 offers a view of the future of reconfigurable antennas and their promise to expand and improve system functionality in a range of different operating scenarios.

·   ·   ·   ·

CHAPTER 2

Definitions of Critical Parameters for Antenna Operation

Traditional characterization of any antenna requires two types of information: the input impedance characteristic over frequency (typically called the frequency response) and the radiation characteristic (or radiation pattern). Usually, frequency response is considered first because without a reasonable input impedance match, a transmitting system may suffer from severe reflections that could damage other components and waste power, whereas receiving systems will suffer from reduced sensitivity and require additional signal amplification. Once an antenna's frequency response is known, the radiation patterns are examined. This chapter briefly reviews both the frequency and radiation characteristics of antennas that can be manipulated through reconfiguration of physical and material parameters as will be shown later.

2.1 FREQUENCY RESPONSE

The frequency response of an antenna is defined as its input impedance over frequency. Complex input impedance (in $Z_{in}(\omega) = R(\omega) + jX(\omega)$ form, with $\omega(= 2\pi f)$ equal to the radian frequency) provides the ability to consider the antenna as a circuit element. As such, the antenna's input impedance can then be used to determine the reflection coefficient (Γ) and related parameters, such as voltage standing wave ratio (VSWR) and return loss (RL), as a function of frequency as given in Eqs. (2.1)–(2.3) [3]:

$$\Gamma = \frac{Z_{in}(\omega) - Z_0}{Z_{in}(\omega) + Z_0} \tag{2.1}$$

$$\text{VSWR} = \frac{V_{max}}{V_{min}} = \frac{1 + |\Gamma|}{1 - |\Gamma|} \tag{2.2}$$

$$\text{RL} = -20\log|\Gamma| \quad (\text{dB}) \tag{2.3}$$

Over the years, antennas have usually (but not always) been designed to have input impedances as close to 50 Ω as possible, which would translate to a reflection coefficient of zero, a VSWR of one, and an RL of infinity in a 50-Ω system. This common practice has supported the standardization of test equipment as well as connectors, cables, and other connections that make assembly of an entire system in a modular fashion straightforward. Keep in mind that if the antenna design is part of a larger system design, there is no requirement that the entire system be based on a 50-Ω reference. Indeed, a different reference point may better suit a power or low-noise amplifier in a number of systems.

Input impedance is usually plotted using a Smith chart [3]. The Smith chart is an indispensable tool for any antenna designer because it shows not only the reflection coefficient but also the nature of the antenna's frequency behavior (inductive or capacitive) that can be used to tune or reconfigure the antenna. Smith chart plots also provide readily accessible information about whether the antenna exhibits resonant behavior. An antenna is resonant at frequencies when its input impedance is purely real; conveniently, this corresponds to locations on the Smith chart where the antenna's impedance locus crosses the real axis.

An antenna's frequency bandwidth is typically defined by the band of frequencies over which its impedance is such that the VSWR at the antenna's input is less than or equal to 2:1 (referenced again to the characteristic impedance of the system, which is usually 50 Ω). This corresponds to an RL that is greater than or equal to 9.54 dB. Narrowband antennas are usually described using a percentage bandwidth, which is calculated as the ratio of the VSWR ≤ 2 : 1 bandwidth to the center frequency of the band. Antennas with wider bandwidths are often specified by the ratio between the upper and lower frequencies that define the bandwidth—such as 3:1 (e.g., for an operating bandwidth between 2 and 6 GHz) or 10:1 (e.g., for an operating band between 100 MHz and 1 GHz). The bandwidth definition can also change depending on different types of antennas or system requirements, but this lecture uses the 50 Ω, VSWR ≤ 2 : 1 bandwidth definition unless otherwise noted.

2.2   RADIATION CHARACTERISTICS

Radiation patterns are graphical representations of an antenna's spatial far-field radiation properties and quantify an antenna's ability to transmit/receive signals in particular directions of interest. They are typically presented as radiation intensity (related to power) or simply electric field (proportional to the square root of the power) in units of decibels (dB) (if the patterns are normalized) or referenced to the gain of an ideal isotropic source (dBi). Radiation plots on linear scales may also be used, but they usually do not capture the full range of interest provided by the logarithmic scale.

The polarization of an antenna's radiation pattern in any particular direction is defined by the figure traced by the extremity of the far-field electric field vector and the sense in which it is traces as observed along the direction of propagation [4]. Polarization is an important property of antennas because it can be used in part to separate or distinguish signals in space, acting as a spatial filter for unwanted signals. There are three kinds of polarization: linear, elliptical, and circular. For elliptical and circular polarizations, the direction of rotation of the field vector is described from the antenna's perspective. That is, with one's thumb pointed in the direction of a propagating wave transmitted from an antenna, right-hand polarization follows the curl of the fingers of the right hand (in a clockwise manner), whereas left-hand polarization follows the curl of the fingers of the left hand (in a counterclockwise manner).

In addition to polarization, antenna radiation patterns may be described using several important parameters that describe their fundamental shapes and characteristics. The first of these is beamwidth defined in a specified two-dimensional plane in the radiating volume of the antenna. The half-power beamwidth of an antenna measures the angle surrounding the direction of maximum radiation across which the antenna's normalized radiation intensity is greater than 0.5. The beamwidth between first nulls is the angle between the first pattern nulls adjacent to the main lobe of the pattern. These two beamwidth descriptors provide information about the shape and extent of the main beam of an antenna or an array of antennas.

Directivity and gain also provide important information about antennas and are especially useful parameters with which to compare competing antenna designs. An antenna's directivity is the ratio of the radiation intensity in a given direction to the radiation intensity averaged over all directions [5]. Gain is the ratio of the radiation intensity, in a given direction, to the radiation intensity that would be obtained if the power accepted by the antenna were radiated isotropically [5]. Gain does not include losses arising from impedance and polarization mismatches. If a direction is not specified for directivity or gain, the direction of maximum radiation intensity is implied. Although the definitions for directivity and gain are very similar, the two parameters are not interchangeable unless the subject antenna is ideal and has no dissipative loss. Because this is rarely the case, a quantity termed the *antenna efficiency*, η, captures the effects of losses and relates an antenna's directivity (D) and gain (G) according to the following equation [6]:

$$G = \eta D \qquad\qquad (2.4)$$

The efficiency of a traditional antenna is typically determined by the ohmic losses created by imperfect conductors and/or dielectrics. In reconfigurable antennas, these kinds of losses may be increased, or other sources of loss may arise. These include losses incurred by any solid-state devices (such as diodes, field-effect transistors [FETs], plasma-based devices, etc.) or other switches or materials used to enable reconfiguration. Reconfigurable antennas based on controlled changes in dielectric or magnetic properties of materials (such as ferroelectrics and ferrites) usually experience more loss because of the presence of often-substantial electrical conductivities of these materials. Effective losses may also be caused by current leakage through control lines or undesired radiation by circuitry used to enable the antenna' reconfiguration. These effects will be discussed in more detail in Chapter 8, which discusses the important practical aspects of reconfigurable antenna realization and operation.

•   •   •   •

CHAPTER 3

Linkage Between Frequency Response and Radiation Characteristics: Implications for Reconfigurable Antennas

Antennas serve as transducers between guided and unguided electromagnetic waves. By bridging the domains of circuits and wave propagation, they have distinct characteristics in both, as described in Chapter 2. The way antennas function as parts of circuits has direct consequences for how they behave as radiators, and vice versa. This, of course, leads to the conclusion that the reconfiguration of one property, say, frequency response, will have an impact on radiation characteristics. Likewise, reconfigurations that result in radiation pattern changes will also alter the antenna's frequency response. This linkage is one of the largest challenges faced by reconfigurable antenna developers who would usually prefer the characteristics to be separable.

One response to this challenge is to apply reconfigurable antennas only where this linkage fits with system requirements or in cases where system requirements can be adjusted to account for the frequency-radiation characteristic linkage. That is, one achieves reconfiguration in one main property of interest and then lives with changes in the others. This is often the case with frequency-reconfigurable or frequency-tunable antennas that have small frequency ranges, which we will examine more closely in the next chapter. In these cases, the small changes in effective antenna length required to change the operating band do not have any significant effects on the radiation pattern of the antenna.

An example of a different kind actually uses the linkage to an advantage, exploiting particular modes of operation to meet specific system requirements. Antennas with multiple resonant modes, for instance, often exhibit very different radiation characteristics for each of the modes. Careful design can allow the bands to be switched or used simultaneously to fully exploit the antenna's innate properties without struggling to suppress or eliminate the frequency response–radiation pattern linkage. One such example of this approach can be found in [7], where the multiband properties of slot spiral antennas can be used to achieve different radiation patterns over different frequency bands for communication in different wireless systems.

In contrast, the third approach is, put bluntly, to struggle. (Most antenna designers will be completely familiar with this approach.) In other words, developers must engage in the struggle to break or condition the link between the antenna's frequency response and radiation characteristics so that one property can be significantly changed independently of the other. For reconfigurable antenna designers in particular and system designers in general, this separation is a holy grail—and something that is obviously not completely attainable for every application. However, in recent years, a number of research groups around the world have taken this approach and achieved very promising results, some of which are discussed in the following chapters. Antennas with these new capabilities have the potential to open up a world of new possibilities for system performance, flexibility, and robustness.

•   •   •   •

CHAPTER 4

Methods for Achieving Frequency Response Reconfigurability

Frequency-reconfigurable antennas (also called tunable antennas) can be classified into two categories: continuous and switched. Continuous frequency-tunable antennas allow for smooth transitions within or between operating bands without jumps. Switched tunable antennas, on the other hand, use some kind of switching mechanism to operate at distinct and/or separated frequency bands. Both kinds of antennas in general share a common theory of operation and reconfiguration—the main differences are in the extent of the effective length changes that enable operation over different frequency bands and the devices and/or means used to achieve these changes. The fundamental theory of operation of these kinds of antennas is presented first, followed by some specific examples and discussions of reconfiguration mechanisms and results for a range of antennas.

4.1   FUNDAMENTAL THEORY OF OPERATION

Many common antennas mentioned in Chapter 1, including linear antennas, loop antennas, slot antennas, and microstrip antennas, are usually operated in resonance. In these cases, the effective electrical length of the antenna largely determines the operating frequency, its associated bandwidth (typically no more than about 10% and usually around 1% to 3% for a single resonance), and the current distribution on the antenna that dictates its radiation pattern. For instance, for a traditional linear dipole antenna, the first resonance occurs at a frequency where the antenna is approximately a half wavelength long, and the resulting current distribution results in an omnidirectional radiation pattern centered on and normal to the antenna axis. In this case, if one wants the antenna to operate at a higher frequency, the antenna can simply be shortened to the correct length corresponding to a half wavelength at the new frequency. The new radiation pattern will have largely the same characteristics as the first because the current distribution is the same relative to a wavelength. The same principle holds true for loops, slots, and microstrip antennas as well.

4.2   RECONFIGURATION MECHANISMS

A number of mechanisms can be used to change the effective length of resonant antennas, although some of these are more effective than others in maintaining the radiating characteristics of the original configuration. The following sections describe different reconfiguration mechanisms, provide some examples, and discuss the benefits and drawbacks of each approach.

4.2.1 Switches

The effective length of the antenna, and hence its operating frequency, can be changed by adding or removing part of the antenna length through electronic, optical, mechanical, or other means. Groups have demonstrated different kinds of switching technology, such as optical switches, PIN diodes, FETs, and radio frequency microelectromechanical system (RF-MEMS) switches, in frequency-tunable monopole and dipole antennas for various frequency bands. For instance, Freeman et al. changed the effective length of a monopole antenna using optical switches, which helped to eliminate some of the switch and bias line effects that can occur with other kinds of switches [8]. A similar design approach was taken by Panagamuwa et al. in [9]. In this case, a balanced dipole fabricated on high-resistivity silicon was equipped with two silicon photoconducting switches, as shown in Figure 4.1 [9]. Light from infrared laser diodes guided with fiber-optic cables was used to control the switches. With both switches closed, the antenna operated at a lower frequency of 2.16 GHz,

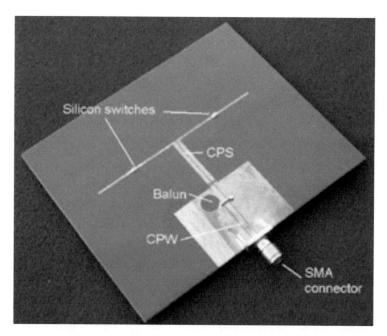

FIGURE 4.1: Photograph of an optically switched dipole antenna that provides frequency reconfigurability (Panagamuwa et al. [9], ©IEEE 2006).

and with both switches open, the antenna operated at 3.15 GHz. The researchers also note variability in antenna gain as a function of the optical power used to activate the switches [9]. A degree of pattern reconfigurability is also achieved with this design by activating only one switch at a time, resulting in a shift of the pattern null in the direction of the longer dipole arm by approximately 50°. However, this pattern reconfiguration is linked to a change in operating frequency, which shifts to 2.7 GHz under these conditions [9]. Kiriazi et al. present a similar example of antenna length changes using RF-MEMS switches in [10]. In this work, opening and closing a pair of RF-MEMS switches reconfigures a simple dipole antenna printed on a high-resistivity silicon substrate to operate in one of two frequency bands.

Using four PIN diodes, Roscoe et al. developed a reconfigurable printed dipole antenna to deliver three operating bands between 5.2 and 5.8 GHz [11]. Others have applied the same approach to microstrip patches [12, 13], microstrip dipoles [2], and Yagi antennas [14]. Radiating structures based on fractal shapes but sharing the same underlying principle have been studied by numerous researchers [e.g., 15–18]. In [15], a three-dimensional fractal tree structure is proposed for use with either passive frequency traps or RF-MEMS switches to deliver operation over multiple frequency bands. Anagnostou et al. discuss a reconfigurable monopole based on a Sierpinski gasket with RF-MEMS switches to connect sections of the antenna together to provide multiple operating bands [16]. Subsequent work with direct integration of RF-MEMS switches provided three separate operating bands with similar omnidirectional radiation characteristics [18]. The simultaneous fabrication of the antenna and switches on a substrate helps to mitigate the effects of package parasitics and other nonideal effects that would be created if the switches were prepackaged and attached using solder and/or bond wires.

Switched frequency-radiating slots with a variety of geometries and radiating properties have also been proposed by a number of researchers. One reconfigurable slot antenna was proposed by Gupta et al. [19]. With a nested ring layout, it was fed with a single slotline or coplanar waveguide (CPW) line. Using eight PIN diode switches, the lower of two operating frequencies was set by the perimeter of the outer loop. When shorter slot sections in two opposing sides of the loop are switched in place, the antenna operates in the upper frequency band [19]. Peroulis et al. [20] demonstrated a tunable antenna using four PIN diode switches that changed the effective length of an S-shaped slot to operate in one of four selectable frequency bands between 530 and 890 MHz. A diagram of the antenna is shown in Figure 4.2 [20]. Changes over such a wide-frequency band are often accompanied with changes in input impedance; however, these investigators positioned the switches and adjusted the slot geometry such that the four frequency bands were attainable through switching alone without needing changes in the matching network or feed point position [20]. Also considered in this design were the parasitic effects and the bias configurations of the PIN diode switches. The DC bias network for each diode switch and its equivalent circuit model are shown

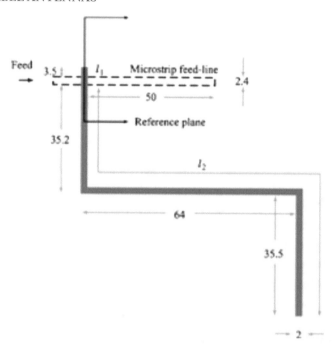

FIGURE 4.2: Diagram of the reconfigurable S-slot antenna, including microstrip feed line. All units are in millimeters. The substrate relative permittivity is 10.2, and the dielectric thickness is 2.54 mm (Peroulis et al. [20], ©IEEE 2005).

in Figure 4.3 [20]. These models were incorporated into simulations to refine the final antenna design.

This brings up another important aspect of reconfigurable antenna design—the compatibility of the antenna topology with the intended reconfiguration mechanism. In some cases, the only way to include a particular switch is to design the antenna around the switch's geometry. One such example of this is the hybrid folded slot/slot dipole antenna described in [21] and shown in Figure 4.4 [21]. This particular design was developed to accommodate a particular RF-MEMS switch layout to simplify the integrated fabrication and operation of the antenna.

One can also achieve discrete changes in an antenna's electrical length by maintaining the active footprint of the antenna but changing the path of the radiating currents on the structure. Yang and Rahmat-Samii demonstrated an example of this approach using a microstrip antenna [22]. A slot is etched in a standard rectangular microstrip patch so that it is perpendicular to the direction of the main current of the patch's first resonance [22]. A PIN diode positioned in the center of the slot changes the current paths on the patch depending on its bias state. With the diode switch

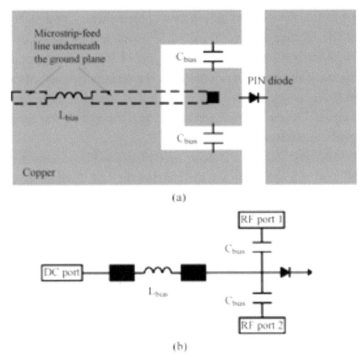

(a)

(b)

FIGURE 4.3: (a) Layout of PIN diode switch bias network for frequency-reconfigurable slot antenna and (b) RF equivalent circuit for bias network (Peroulis et al. [20], ©IEEE 2005).

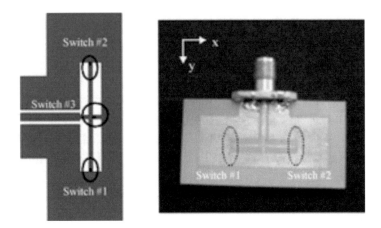

FIGURE 4.4: Reconfigurable hybrid folded slot/slot dipole antenna: antenna geometry including all three switch locations (left) and photograph of the fabricated antenna using only two of the three switches (right) (Huff and Bernhard [21], ©IEEE 2005).

open, currents travel around the slot and the antenna operates in a lower frequency band. With the diode switch closed, the effective length of the patch is shorter and the antenna operates in a higher frequency band. The slot length controls the frequency ratio between the upper and lower operating frequencies. As long as the slot length is not too long, the radiation pattern of the original antenna is largely preserved [22]. Longer slot lengths will result in radiation patterns with higher cross-polarization components. Others have since extended this concept to other microstrip structures, such as E-shaped patch antennas [23].

Although these are just some of the many examples of switched frequency reconfigurability with a range of different antenna types and geometries, they all share the common approach of discrete changes in effective length to achieve their goals.

4.2.2　Variable Reactive Loading

The use of variable reactive loading has much in common with the switched reconfigurability discussed in the previous chapter. The only real difference between the two is that, in this case, the change in effective antenna length is achieved with devices or mechanisms that can take on a continuous range of values (typically capacitance) that allows smooth rather than discrete changes in the antenna's operating frequency band.

One example of a continuously tuned microstrip patch antenna is presented in [24]. In this case, two varactor diodes (varactors) were connected between the main radiating edges of the structure and the ground plane. With a reverse bias varying between 0 and 30 V, the varactors had capacitances between 2.4 and 0.4 pF. As the bias level changed, the capacitances at the edges of the patch tuned the effective electrical length of the patch. Continuous frequency tuning over a large band is possible (20–30% as shown in [24]) depending on the antenna topology.

A one-wavelength slot antenna loaded with two one-port reactive FET components was tuned continuously in [25]. By changing the bias voltage, the reactances of the FETs were varied by changing the bias voltage, which, in turn, changed the effective length of the slot and its operating frequency. The range of tuning was about 10%, centered around 10 GHz. The patterns were essentially unchanged for this relatively small tuning range [25]. Similar tunable slot antennas equipped with varactors [e.g., 26, 27] have also been developed, which take advantage of higher-order resonances to create tunable dual-band performance. Using a transmission line model of the loaded slot resonator, the varactors' positions can be determined to enable independent tuning of the two bands [27].

More recently, a microstrip patch antenna has been tuned using integrated RF-MEMS capacitors [28]. Shown in Figure 4.5 [28], the capacitors are implemented on a CPW tuning stub and actuated with continuous DC bias voltages up to 12 V, which produce operating frequencies

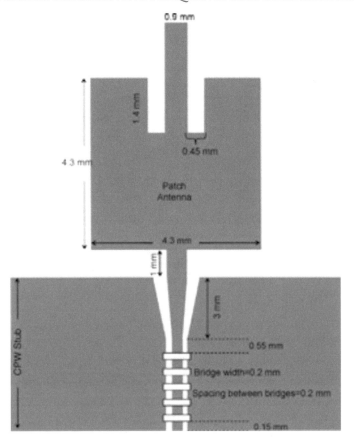

FIGURE 4.5: Frequency-tunable microstrip patch antenna with RF-MEMS capacitors and CPW tuning stub (Erdil et al. [28], ©IEEE 2007).

between 15.75 and 16.05 GHz [28]. The unique monolithic approach to the design eliminates the need for bias vias, and the resulting radiation patterns show little effect of the proximity of the tuning stub. Other microstrip antennas with slots equipped with solid-state varactors have also been demonstrated [29].

A combination of switching and reactive tuning has also been implemented to support both course and fine frequency tuning in a printed monopole antenna [30]. Based on a meander-line monopole structure, a PIN diode is implemented to provide course tuning between system bands (in this case, 2 and 5 GHz), whereas a varactor is used for fine tuning within each band [30], perhaps supporting a level of signal filtering capability at the antenna that can be used in conjunction with other radio functions.

FIGURE 4.6: Photograph of mechanically actuated reconfigurable antenna with movable parasitic element, providing variable operating frequency, bandwidth, and gain (Bernhard et al. [33], ©IEEE 2001).

4.2.3 Structural/Mechanical Changes

Mechanical rather than electrical changes in antenna structure can deliver larger frequency shifts, whether used for switched or continuously variable bands. The main challenges with these antennas lie in the physical design of the antenna, the actuation mechanism, and the maintenance of other characteristics in the face of significant structural changes. One example of a mechanically tuned antenna was demonstrated in 1998, where a piezoelectric actuator system was used to vary the spacing between a microstrip antenna and a parasitic radiator to change the operating frequency of the antenna [31–33]. A picture of the antenna is presented in Figure 4.6 [33]. Although normally possessing a very narrow bandwidth (1%), controlled movement of the parasitic element delivered an effective bandwidth of about 9%. This example illustrates the difficulty in achieving one kind of reconfigurability without incurring changes in other antenna characteristics; the bandwidth and gain of the structure also change as a function of parasitic element spacing but cannot be individually selected [33].

Another example of continuous frequency changes enabled by mechanical changes is a magnetically actuated microstrip antenna [34]. A microstrip antenna designed for operation around 26 GHz was covered with a thin layer of magnetic material and released from the substrate. Using a micromachining process called plastic deformation assembly, application of an external DC magnetic field causes plastic deformation of the antenna at the boundary point where it is attached to the microstrip feed line, resulting in a patch positioned at an angle over the substrate. A photograph

FIGURE 4.7: Photograph of magnetically actuated reconfigurable microstrip antenna (Langer et al. [34], ©IEEE 2003).

of one prototype is given in Figure 4.7 [34]. Small changes of the angle at which the structure resides results in changes in operating frequency that preserve radiation characteristics, whereas larger angles result in frequency shifts accompanied by significant changes in the antenna's radiation pattern. In particular, as the elevation angle between the patch and the horizontal substrate increases past 45°, the antenna's radiation is more characteristic of a horn antenna and changes toward the pattern of a monopole antenna as the angle approaches 90°.

4.2.4   Material Changes

Although changes to the conductors predominate in reconfigurable antenna designs, changes in the material characteristics of designs also promise the ability to tune antennas in frequency. In particular, an applied static electric field can be used to change the relative permittivity of a ferroelectric material, and an applied static magnetic field can be used to change the relative permeability of a ferrite. These changes in relative permittivity or permeability can then be used to change the effective electrical length of antennas, again resulting in shifts in operating frequency. As a potential bonus, their relative permittivities and permeabilities are high compared with commonly used substrate materials, translating into greatly reduced antenna sizes. Aside from any complexities resulting from the necessity of the bias structure, the main drawbacks to using standard ferroelectric and ferrite bulk materials (typically with thicknesses on the order of millimeters) are their high-conductivity relative to other substrates that can severely degrade the efficiency of the antenna.

One example of a frequency-tuned ferrite-based antenna is presented in [35], which provided a 40% continuous tuning range with the variable static magnetic field in the plane of the substrate and perpendicular to the resonant dimension of the patch. However, the radiation performance of the design left much to be desired, with cross-polarization levels that were significantly higher than those expected from a traditional rectangular microstrip antenna [35]. Others have also investigated the properties of ferrite-based microstrip antennas [e.g., 36, 37], with results indicating that factors including nonuniform bias fields and the multiple modal field distributions excited in a bulk ferrite substrate may preclude their use in practical applications.

Recently, several groups have developed ferroelectric materials in thin film form in an effort to minimize the loss introduced into the circuit while still providing a degree of tunability [e.g., 38–40]. However, most proposed applications still use tunable materials in the feed structure or in parasitic elements rather than the antenna itself due to limitations in the planar extent and achievable uniformity of the films.

• • • •

CHAPTER 5

Methods for Achieving Polarization Reconfigurability

Antenna polarization reconfiguration can help provide immunity to interfering signals in varying environments as well as provide an additional degree of freedom to improve link quality as a form of switched antenna diversity [41]. They can also be used in active read/write tracking/tagging applications [41].

5.1 FUNDAMENTAL THEORY OF OPERATION

The direction of current flow on the antenna translates directly into the polarization of the electric field in the far field of the antenna. To achieve polarization reconfigurability, the antenna structure, material properties, and/or feed configuration have to change in ways that alter the way current flows on the antenna. Polarization reconfigurations can take place between different kinds of linear polarization, between right- and left-handed circular polarizations, or between linear and circular polarizations. The mechanisms to achieve these modifications (e.g., switches, structural changes) are largely the same as those described for frequency reconfigurability earlier, although their implementations are necessarily different. The main difficulty of this kind of reconfigurability is that this must be accomplished without significant changes in impedance or frequency characteristics.

5.2 RECONFIGURATION MECHANISMS

5.2.1 Switches

Several antennas have been developed to deliver reconfigurable polarization characteristics using switches. One example of such a polarization-agile antenna is the "patch antenna with switchable slots," or PASS antenna [42, 43], which has also been used to provide frequency reconfigurability as discussed previously [22]. In general, the PASS antenna consists of a microstrip antenna with one or more slots cut out of the copper patch. A switch (either solid-state or RF-MEMS) is inserted in the center of the slot to control how currents on the patch behave. When the switch is open, currents must flow around the slot. When the switch is closed, the current can follow the shorter path created by the closed switch. Polarization reconfigurability is achieved by including two orthogonal slots on the surface of the patch. Alternate activation of the switches yields either right- or left-hand circular polarization. Others patch antennas with switchable circular polarization, using switches in the feed excitation slots rather than the surface of the patch, have also been proposed for active read/write microwave tagging systems [41]. Patches with switched corners for circular polarization control have also been presented [44].

Slot antennas such as the one developed in [45] have also been implemented to deliver polarization reconfigurability. Fries et al. describe a slot-ring antenna equipped with PIN diodes to reconfigure between linear and circular polarization or between two circular polarizations [45]. Figure 5.1 [45] shows the basic topology of the antenna, whereas Figure 5.2 [45] depicts the specific diode positions, biasing, and ground plane configurations for both designs [45]. For the linear-circular design (Figure 5.2a), forward biasing the diodes across the small discontinuities at 45° and –135° relative to the feed point delivers linear polarization, whereas reverse biasing the diodes results in circular polarization [45]. The design in Figure 5.2b includes additional symmetric discontinuities to support switching between left- and right-handed circular polarizations [45]. In both designs, the ground planes are carefully designed to support proper DC biasing for the diodes while providing RF continuity through capacitors connected between ground plane sections. This radiating structure is a good example of the additional factors that must be taken into account when transitioning from a fixed to a reconfigurable antenna—the fundamental structure may remain the same, but critical adjustments are required to enable proper DC connections and RF performance.

Another polarization-reconfigurable antenna uses a MEMS actuator, which can be considered to be a kind of switch [46]. In this design (shown in Figure 5.3 [46]), the MEMS actuator is

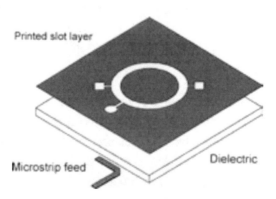

FIGURE 5.1: Basic topology of a microstrip-fed circularly polarized slot ring antenna (Fries et al. [45], ©IEEE 2003).

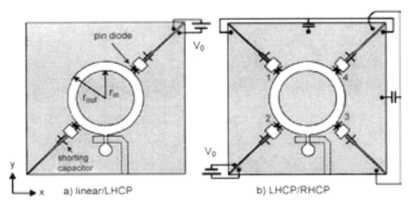

FIGURE 5.2: Two polarization-reconfigurable slot ring antennas: (a) switchable between linear and left-hand circular polarization and (b) switchable between left- and right-hand circular polarization (Fries et al. [45], ©IEEE 2003).

located within a simple microstrip patch antenna designed to support two orthogonal modes when excited in the corner. The actuator consists of a moveable metal strip suspended over a metal stub. When the strip is suspended above the stub, the antenna has a circularly polarized radiation pattern [46]. Using electrostatic actuation, the metal strip can be lowered to create an antenna with dual linear polarization. A related structure that uses piezoelectric transducers to achieve switching between left- and right-hand circular polarization with a microstrip patch antenna is shown in Figure 5.4 [47].

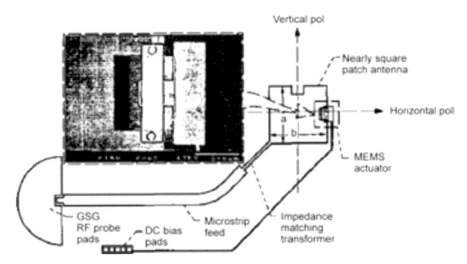

FIGURE 5.3: Polarization-reconfigurable microstrip patch antenna using an integrated MEMS actuator. Inset shows photomicrograph of the MEMS actuator (Simons et al. [46], ©IEEE 2002).

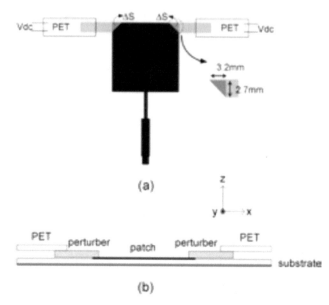

FIGURE 5.4: (a) Top and (b) side views of a polarization-reconfigurable microstrip antenna equipped with piezoelectric transducers (PET) (Hsu and Chang [47], ©IEEE 2007).

5.2.2 Material Changes

Rainville and Harackiewicz [48] presented a polarization-tunable microstrip antenna based on static magnetic biasing of a ferrite film. As with microstrip antennas on bulk ferrite substrates [e.g., 35–37], the copolarized fields are much less dominant than those from a traditional microstrip patch antenna. The authors took advantage of the fact that the applied static bias field tuned the frequency of the cross-polarized field to create a range of elliptical polarizations. Optimization of feed point and ferrite film properties could result in purely circular and linear polarizations as well [48].

•   •   •   •

Methods for Achieving Radiation Pattern Reconfigurability

This chapter details a number of unique approaches to pattern-reconfigurable antenna design, most of which preserve frequency characteristics. Antenna radiation is typically characterized with a set of two-dimensional radiation patterns taken from major cut planes of interest, with three-dimensional patterns used occasionally to obtain a qualitative assessment of the antenna's performance. In this lecture, we will typically examine cut planes of radiation patterns for clarity and simplicity.

6.1   FUNDAMENTAL THEORY OF OPERATION

The arrangement of currents, either electric or magnetic, on an antenna structure directly determines the spatial distribution of radiation from the structure. This relationship between the source currents and the resulting radiation makes pattern reconfigurability without significant changes in operating frequency difficult, but not impossible, to achieve. To develop antennas with specific reconfigurable radiation patterns, a designer must determine what kinds of source current distributions, including both magnitude and phase information, are necessary. Once a topology for the current distribution is determined, a baseline antenna design can be selected and then altered to achieve the desired source current distribution. This design process is very much akin to that of array synthesis. The remaining task is to either arrange the design so that the frequency characteristics are largely unchanged or to compensate for changes in impedance with tunable matching circuits at the antenna terminals. In some cases, antenna types (such as reflector antennas or parasitically coupled antennas) are selected so that the input is more isolated from the reconfigured portion of the structure, allowing the frequency characteristics to remain relatively unchanged while radiation patterns are reconfigured.

It should be noted that some antennas have been developed to deliver multiple radiation patterns at a single frequency through careful design of feed structures to couple degenerate modes that possess different radiation patterns [e.g., 49]. Although these kinds of designs do not strictly meet the definition of reconfigurability described in Chapter 1, they do provide insight into a general approach for developing such designs.

6.2 RECONFIGURATION MECHANISMS

6.2.1 Structural/Mechanical Changes

With the reflective surface physically removed and isolated from the primary feed, reflector antennas are a natural choice for applications that require radiation pattern reconfiguration independent of frequency. Clarricoats and Zhou demonstrated an example of a radiation-reconfigurable reflector antenna by actively changing the structure of a mesh reflector [50]. In its first embodiment, the reflector contour was changed manually in certain regions, which resulted in changes in beam shape and direction [50]. Later, computer-controlled stepper motors were implemented to pull cables attached to specific points on the reflector mesh to support automatic pattern reconfiguration [51].

More recently, a similar system for satellite applications has been developed with expanded capabilities enabled by changes in the system's subreflector rather than the main reflector [52]. Shown in Figure 6.1 [52], the subreflector is fabricated using a thin, flexible, conductive material, and its shape is changed using piezoelectric actuators attached to its back surface. A desired radiation pattern is achieved when the actuators deform the subreflector surface and produce changes in the electromagnetic field illuminating the main reflector. To minimize the number of actuators (and, hence, system complexity and weight), their positions are determined using an iterative finite-element algorithm. This algorithm indicates where successive actuators should be placed to minimize the error between the desired and actual subreflector shape [52]. Figures 6.2 and 6.3 show the beam patterns supported by the antenna reconfiguration [52].

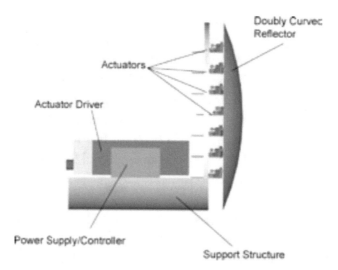

FIGURE 6.1: Reconfigurable subreflector system, including piezoelectric actuators and driver to deflect the surface of a Cassegrain subreflector (Washington et al. [52], ©IEEE 2002).

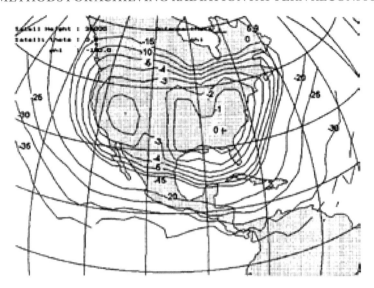

FIGURE 6.2: Radiation pattern for U.S. coverage using reflector antenna equipped with the mechanically reconfigurable subreflector (Washington et al. [52], ©IEEE 2002).

Changes in an effective reflector have also been demonstrated using reconfigurable high-impedance surfaces [53]. Here, a lattice of small resonant elements produces high-surface impedance near their resonant frequency, creating, in effect, an artificial ground plane. Changing the capacitances between resonant elements across the lattice through mechanical means creates a phase gradient that can produce a steered beam when the lattice is used as a reflector. An electrically tuned version of this structure using varactors that can deliver two-dimensional beam steering is discussed in [54].

A closely related approach has also been used to develop a reconfigurable leaky-wave antenna using mechanical tuning [55]. In this case, a horizontally polarized antenna is used to couple energy into leaky transverse electric waves on a tunable impedance surface. A diagram of the surface, including the moveable top capacitance surface, is shown in Figure 6.4 [55]. The radiated beam from the surface can be steered in elevation over a range of 45° by changing the apparent capacitance of the surface through mechanically shifting the top capacitive plane. The antenna and the surface are shown in Figure 6.5 [55]. An electrically tuned version of this antenna that uses varactors can produce reconfigurable backward as well as forward leaky-wave beams [56, 57].

The electromechanically scannable trough waveguide antenna, first developed in 1960, is one early example of a pattern-reconfigurable antenna [58]. Based on an asymmetrical trough waveguide intended to feed a parabolic reflector [59], two different mechanisms were proposed to alter the phase velocity in the guide that resulted in scanned radiation patterns at a single frequency. The first mechanism involved the rotation of artificial dielectric structures along a longitudinal axis

FIGURE 6.3: Radiation pattern for Brazil coverage using reflector antenna equipped with the mechanically reconfigurable subreflector (Washington et al. [52], ©IEEE 2002).

within the trough waveguide, and the second used a mechanical variation of the height of periodic structures located on the top of the center fin of the trough waveguide [58]. Although elegant in concept, the authors acknowledged at the time that both approaches presented major mechanical issues that they resolved using "cam-and-gear" solutions that were complex and cumbersome.

Huff and Bernhard [60] began with the concept of [58] and proposed the application of more modern perturbations and streamlined actuation mechanisms to achieve beam steering at a single frequency while preserving the wide bandwidth characteristics of the original antenna. The trough waveguide's fundamental structure is shown in Figure 6.6 [60]. In the more modern embodiment, beam steering is achieved using micromachined cantilever structures fabricated within the trough waveguide, shown in Figure 6.7 [60]. These cantilevers, which can be actuated electrostatically or magnetostatically, create perturbations that change the apparent phase velocity in the waveguide and produce scanned radiation patterns. Simulations with actuation of individual cantilevers or groups of cantilevers demonstrate beam scanning at single frequencies through broadside for this traveling wave antenna [60].

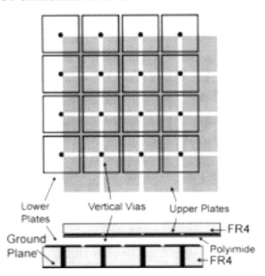

FIGURE 6.4: Mechanically reconfigurable impedance surface consisting of two printed circuit boards: a high-impedance ground plane and a separate tuning layer. The tuning layer is moved across the stationary high-impedance surface to vary the capacitance between the overlapping plates and tune the resonance frequency of the surface (Sievenpiper et al. [55], ©IEEE 2002).

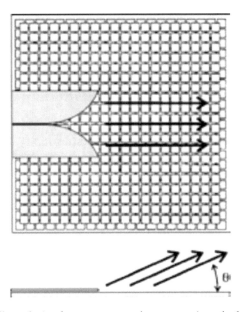

FIGURE 6.5: A horizontally polarized antenna couples energy into leaky modes on the tunable impedance surface. The waves propagate across the surface and radiate at an angle governed by the surface resonance frequency with respect to the excitation frequency. By tuning the surface resonance frequency, the beam is steered in the elevation plane (Sievenpiper et al. [55], ©IEEE 2002).

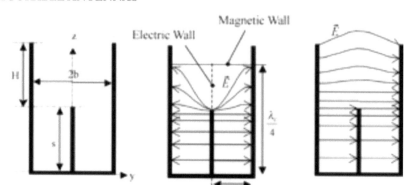

FIGURE 6.6: Trough waveguide dimensions (left), electric field lines of the dominant transverse electric mode with equivalent structure for design and field distribution (middle), electric field lines of the transverse electromagnetic mode (right) (Huff and Bernhard [60], ©Huff and Bernhard 2005).

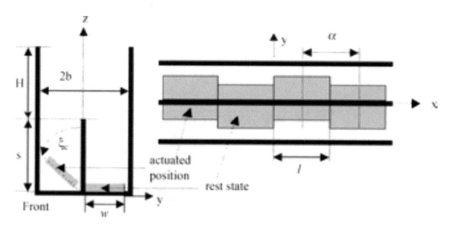

FIGURE 6.7: Trough waveguide with cantilever-type perturbations, showing staggered positions of actuated and rest states of the cantilevers (Huff and Bernhard [60], ©Huff and Bernhard 2005).

Another approach using micromachined components has also been demonstrated that uses mechanical changes in a hornlike antenna to produce reconfigurable radiation patterns with no appreciable changes in frequency response [61]. The basic antenna, shown in Figure 6.8 [61] is composed of a planar "V" structure with a coplanar transmission line feed. Micromachined rotational hinges fixed to the substrate material hold the ends of each of the "V" arms in place. MEMS actuators move the arms laterally to alter the radiated main beam direction and/or beam shape [61]. Figures 6.9 and 6.10 [61] show some of the possible beam variation achieved with this structure [61].

Chang et al. enable mechanical perturbation of propagation constants in a dielectric waveguide antenna operating at millimeter-wave frequencies to achieve pattern reconfigurability [62].

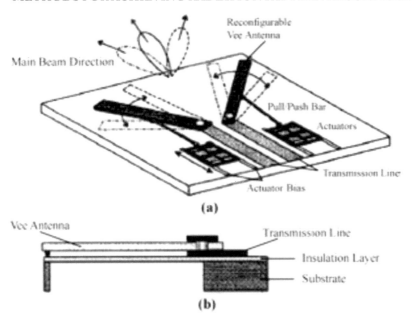

FIGURE 6.8: (a) Conceptual diagram and (b) cross section of a MEMS reconfigurable Vee antenna (Chiao et al. [61], ©IEEE 1999).

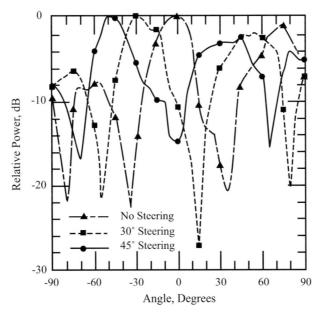

FIGURE 6.9: E-plane beam-steering patterns for a 17.5-GHz 75° Vee antenna (Chiao et al. [61], ©IEEE 1999).

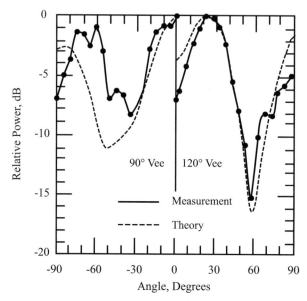

FIGURE 6.10: Measured and theoretical E-plane patterns (in halves) for 90° and 120° Vee antennas showing reconfigurability in beam forming (Chiao et al. [61], ©IEEE 1999).

The propagation constant along a dielectric image line is gradually perturbed with a thin, moveable film placed on top of it. The antenna is shown in Figure 6.11 [62]. Different positions of the grating film present different apparent grating spacings to the traveling wave and result in a scanned beam. Scan angles of up to 53° have been demonstrated with this design at 35 GHz, with lower scan angles achievable over an operating band between 30 and 40 GHz [62]. The achievable beam steering at three frequencies is shown in Figure 6.12 [62]. Scanned dual beam performance over a similar frequency range has been achieved with a related but more complex structure [63].

6.2.2 Electrical Changes

In general, electrical changes to a radiating structure usually result in changes in radiation characteristics. Knowledge of the fundamental operation of the antenna in question can help to make the appropriate changes to arrive at a useful design. For example, reconfigurable radiation patterns can be achieved with slot-based radiators. In [64], an annular slot antenna is used as both a frequency- and pattern-reconfigurable device. Frequency reconfigurability for this antenna is supported through PIN diode switches that control input matching circuitry, whereas the pattern reconfigurability is enabled with diode switches placed at locations around the slot to control the direction of a pattern null that is inherent to basic antenna operation [64].

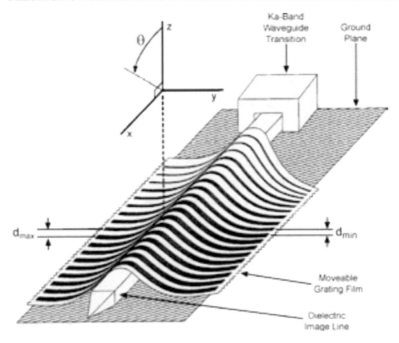

FIGURE 6.11: A traveling wave antenna based on moveable grating fed by a dielectric image line, capable of beam scanning at millimeter-wave frequencies (Chang et al. [62], ©IEEE 2002).

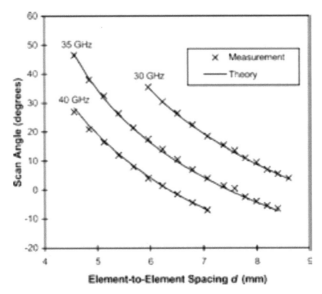

FIGURE 6.12: Measured and calculated beam scanning along the θ direction at 30, 35, and 40 GHz as a function of element-to-element spacing, d, in the moveable grating (Chang et al. [62], ©IEEE 2002).

6.2.2.1 Parasitic Tuning

One of the most effective and widespread methods to change radiation patterns independently from frequency behavior is the use of electrically tuned or switched parasitic elements. These methods possess several attractive qualities: isolation of the driven element(s) from the tuned element(s), potentially wide-frequency bandwidth, and a range of available topologies and functionalities. Fundamentally, tuning of antenna radiation patterns in this manner relies on the mutual coupling between closely spaced driven and parasitic elements, resulting in effective array behavior from a single feed point. Therefore, changes in radiation patterns are achieved through changes in the coupling between the elements, which, in turn, change the effective source currents on both the driven and parasitic elements. Relative leads or lags in the induced current on the parasitic element(s) result in classical "reflector" or "director" behavior that leads to steered or tilted beams.

In 1978, Harrington proposed a parasitic dipole array that continues to see application today in various related forms [65]. Figure 6.13 [65] shows the driven dipole element surrounded by parasitic dipoles loaded with tunable reactances. Variations in the loading reactance of each parasitic element changes the apparent magnitude and phase of the signal on each array element, resulting in a directive beam in a desired direction [65]. Luzwick and Harrington also proposed a waveguide-based reconfigurable array based on the same concept [66]. In the intervening years, a number of reconfigurable parasitic arrays have been proposed and studied, using both switched and reactively loaded elements, which provide a wide variety of functionality [e.g., 67–74]. Note that although it is easier to achieve changes in beam steering and forming with a parasitic array, the coupling between the driven element and the loaded parasitic elements can still affect the input impedance of the antenna [70].

Microstrip-based reconfigurable antennas can also use switched or tuned parasitic elements. One example is that developed by Zhang et al. [75]. Shown in Figure 6.14 [75], the antenna is composed of a single linear element with two spaced parasitic elements positioned parallel to the

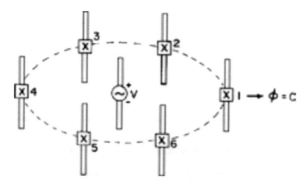

FIGURE 6.13: A seven-element circular array of reactively loaded parasitic dipoles for reconfigurable beam steering and beam forming (Harrington [65], ©IEEE 1978).

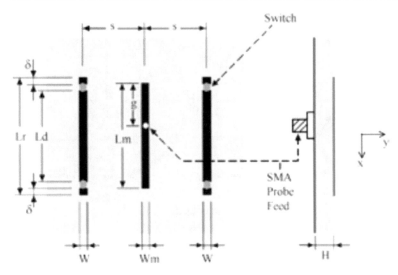

FIGURE 6.14: Physical structure and parameters of the reconfigurable microstrip parasitic array. Switching or tuning of parasitic elements on either side of the driven element provides beam tilt capabilities over a common impedance bandwidth (Zhang et al. [75], ©IEEE 2004).

driven element. Parasitic element lengths are changed with electronic switches [75] or varactors [76], which, in turn, alter the magnitudes and phases of the currents on the parasitic elements relative to the driven element. Tilts in the main beam in one plane can then be switched [75] or swept [76] as the lengths of the parasitic elements are changed. A photograph of the antenna is shown in Figure 6.15 equipped with PIN diode switches [77], with its switched beams shown in Figure 6.16 [75]. Note the careful design of the switch control circuitry shown in Figure 6.15 that minimizes

FIGURE 6.15: Photograph of the reconfigurable microstrip parasitic array (Zhang [77], ©Zhang 2005).

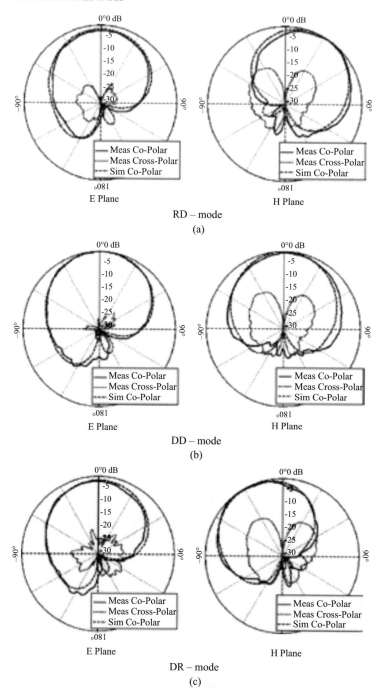

FIGURE 6.16: Reconfigured beam patterns from the antenna shown in Figure 6.15: (a) reflector-director (RD), (b) director-director (DD), (c) director-reflector (DR) (Zhang et al. [75], ©IEEE 2004).

the effects of radiation from DC bias lines [77]. With the driven element relatively isolated from the reconfigured sections of the structure, the operating frequency and impedance bandwidth are preserved. This antenna, in a similar manner to most parasitically tuned antennas, can be analyzed theoretically using a combination of coupling and array theory to explain pattern tilts [77]. Other examples of small reactively steered planar arrays based on standard microstrip patches are provided by Dinger [78, 79]. For tuned parasitic systems such as these, search and optimization algorithms can be used to determine the tuning reactances necessary on each parasitic element to produce a beam or null at a prescribed angle [73, 78]. However, the close spacing between the driven and parasitic elements in some cases may also affect the impedance bandwidth of the system.

Although the parasitic reconfigurable antennas discussed thus far have been relatively straightforward in their topology and operation, other more complex structures have been developed that operate on the same principle but are more difficult to analyze and design. One example of this is the reconfigurable square microstrip spiral developed by Huff et al. [80] that provides a broadside or 45° tilted beam over a common impedance bandwidth shown in Figure 6.17. The operating band is comparable to that of a microstrip patch on an electrically thin substrate. In its fundamental state as a simple spiral microstrip antenna, it delivers a broadside radiation pattern. The antenna has two switched connections: one that shorts the end of the spiral to ground (shown on the left of Figure 6.17 [80]) and one that opens a small gap in the spiral arm (shown on the right of Figure 6.17 [80])]). When the two switches are activated, in this manner the antenna becomes, in essence, an open microstrip line with a parasitic arm (formed by the end section of the spiral formed by the open line that is now shorted to ground at one end). In this configuration, a 45° tilt from broadside in radiation pattern results. Opening the gap in the line without shorting the end of the spiral also results in a broadside radiation pattern at a higher frequency [80]. Antennas based on a similar structure have also delivered switched broadside and end-fire radiation characteristics over a common impedance bandwidth with RF-MEMS–switched connections [81–83]. Related microstrip spiral-based designs that can deliver tilted beam patterns by virtue of the tuning of effective parasitics in the outer spiral structure have also been studied by others [e.g., 84, 85].

6.2.2.2 Array Tuning

Integrating phase shifts into array element reconfigurability can result in beam steering similar to that achieved with traditional phased arrays but without the inherent costs of phase shifters. Several reconfigurable apertures and surfaces based on this principle are discussed here.

One approach to achieve radiation pattern reconfiguration uses phased-tuned reflectarray elements [86]. In this work, an electronically scanned reflectarray uses reconfigurable microstrip patch antenna elements to vary the reflection phase across the array. The reconfigurable element,

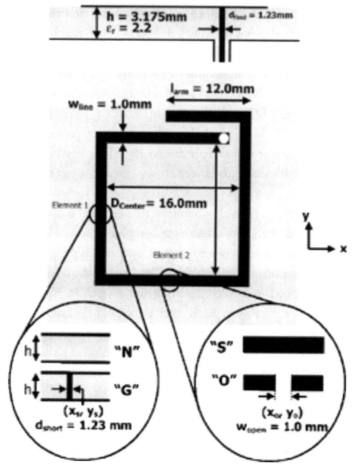

FIGURE 6.17: Radiation pattern- and frequency-reconfigurable square spiral microstrip antenna showing switch placements necessary for reconfiguration (Huff et al. [80], ©IEEE 2003).

shown in Figure 6.18 [86], is a simple microstrip patch element with aperture coupling to a transmission line loaded with two varactor diodes. By varying the bias across the two varactors, each element reflection phase can be varied over 360°, which supports array beam steering to up to 40° from broadside with a 30-element array [86].

6.2.3   Material Changes

Ferrites and ferroelectric materials, although typically applied in frequency reconfigurability, can also be used to reconfigure radiation patterns. In these cases, the changes in material characteristics can be used to change the resonant current distributions on conductors, which then result in radia-

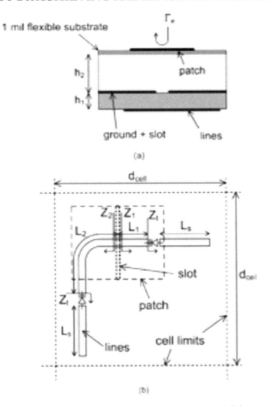

FIGURE 6.18: Reconfigurable reflectarray element (a) side view and (b) bottom view (Riel and Laurin [86], ©IEEE 2007).

tion pattern changes, or they can be used to alter propagation speeds in traveling/leaky-wave radiators that result in beam steering.

During the past 60 years, several researchers have investigated the use of the tunability of ferrite materials to produce steerable radiation patterns [87]. In [87], a ferrite superstrate positioned above a circular microstrip patch antenna provided up to 15° of beam tilt when the ferrite was biased with a permanent magnet. However, the losses and required thickness of the ferrite prevented further practical implementation.

In [88], the variable permittivity of a ferroelectric superstrate is used to achieve frequency-fixed beam steering with a two-dimensional grid array of resonant slot antennas. Shown in Figure 6.19 [88], the complete structure consists of a grounded nonmagnetic substrate with a tunable ferroelectric superstrate, which is then covered with a conducting plate that supports a two-dimensional array of radiating slots. When fed with an RF signal from below, changes to the applied bias between the conducting plate and the ground plane tune the permittivity of the

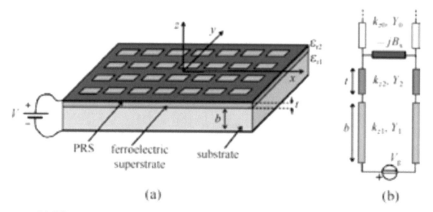

(a)   (b)

FIGURE 6.19: (a) Three-dimensional view of the ferroelectric reconfigurable leaky-wave antenna, with the relevant physical and geometrical parameters. (b) Transverse equivalent network model of the antenna in (a), excited by a magnetic dipole on the ground plane (Lovat et al. [88], ©IEEE 2006).

ferroelectric and the beam direction of the structure changes [88]. Although the fundamental operation of this antenna was studied using numerical techniques, the practicality of achieving required bias voltages for ferroelectric tuning in this kind of parallel plate tuning configuration still remains an issue. Other reconfigurable leaky-wave antennas that include tunable ferroelectric materials in planar configurations have also been reported [e.g., 89, 90].

A two-dimensional beam steering antenna based on a combination of ferroelectric material tuning and the continuous transverse stub topology has been demonstrated [91]. Similar in conceptual approach to other reconfigurable leaky-wave antennas that possess integrated phase shifting, ferroelectric material integrated into the feeding waveguide is tuned to deliver desired phase shifts between radiating stubs that then results in beam steering up to 60° from broadside. A diagram of the basic antenna structure is shown in Figure 6.20 [91]. Interestingly, the height of the waveguide (and hence, the thickness of the loading ferroelectric material) needs to be varied across the array to preserve good impedance characteristics. This need arises because as the permittivity of the ferroelectric loading material is varied with applied bias, the electrical distance between the radiating stubs and the characteristic impedance of the waveguide change [91]. An antenna based on similar concepts but using individually biased ferrite rods between radiating slots in a waveguide is discussed in [92].

Ferrites have also been implemented in leaky-wave antenna structures to achieve beam scanning. In [93], a corrugated ferrite (yttrium iron garnet) slab supported by a dielectric waveguide (shown in Figure 6.21 [93]) provides continuous beam scans up to 40° from broadside at 46.8 GHz with changes in applied dc magnetic field. A thorough comparison of experimental results with theoretically driven predictions of performance points to the need for further development of fer-

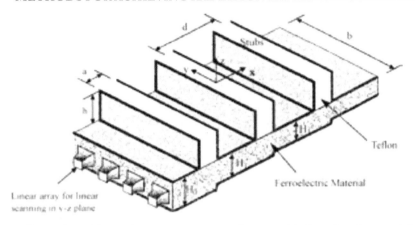

FIGURE 6.20: Reconfigurable continuous transverse stub array with ferroelectric material fill for beam scanning in the x–z plane (Iskander et al. [91], ©IEEE 2001).

rites with larger saturation magnetizations [93]. More conventional approaches have used biased ferrites in phase shifters to achieve beam steering in arrays [e.g., 94–96], but these structures do not strictly qualify as reconfigurable antennas because the antennas' structures themselves remain unchanged and beam steering relies strictly on traditional phased array techniques.

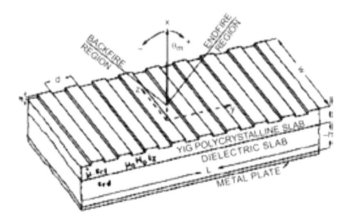

FIGURE 6.21: Geometry of the reconfigurable/tunable corrugated ferrite slab/dielectric layer structure (Maheri et al. [93], ©IEEE 1988).

CHAPTER 7

Methods for Achieving Compound Reconfigurable Antennas

As illustrated in the previous chapters, it is extremely challenging to separate an antenna's frequency characteristics from its radiation characteristics. Indeed, this ability to independently select operating frequency, bandwidth, and radiation pattern characteristics is the ultimate goal of reconfigurable antenna designers. Recently, several groups have achieved this kind of reconfigurability, here termed *compound reconfigurability*.

A few of the antennas already discussed are already capable of separate and selectable impedance and radiation performance, (e.g., [64, 80]). In [64], an annular slot antenna is used as both a frequency- and pattern-reconfigurable device. In [80], a resonant single-turn microstrip spiral antenna provides broadside radiation operation at two different frequencies or end-fire radiation characteristics at one of these frequencies. A Yagi-based approach using reconfigurable slot-loaded parasitic elements as directors and reflectors has been presented that supports tilted or broadside beams at two separate frequencies [97]. In [98], a reconfigurable stacked microstrip antenna delivers a broadside circularly polarized beam at one frequency and a dipole pattern at a lower frequency for joint satellite and terrestrial operation.

More extreme embodiments that adopt novel approaches to deliver "general-purpose" reconfigurable apertures typically require the use of optimization, genetic, or other algorithms to arrive at a final configuration of multiply-connected subwavelength conductive regions that delivers specified function.

7.1   FUNDAMENTAL THEORY OF OPERATION

Although the fundamental theory of operation behind any compound reconfigurable antenna is no different from that of an ordinary reconfigurable antenna, design and control in this case are obviously more complex. To date, most single-element compound reconfigurable antennas have a primary focus on one operating dimension and then additional functionality is achieved by tuning of the existing structure. However, in the case of reconfigurable apertures, the frequency and radiation characteristics are determined together.

A general purpose reconfigurable aperture, geared to serve extremely wide bandwidths (of ratios of 10:1 or more in some cases), is motivated by the desire to have a flexible palette on which to establish current or field distributions that result in desired antenna behavior. In essence, these apertures are intended to create whatever antenna is necessary to achieve specific frequency, bandwidth, radiation, and polarization characteristics. Researchers have developed pixel-based approaches that allow them to divide up the aperture into small parts and use this granularity to control the current or field distributions in the aperture. For most desired aperture functions, how the aperture should be configured or fed is not immediately apparent, so computational algorithms have been applied to arrive at suitable and sometimes optimal designs.

7.2   ELECTRONIC RECONFIGURATION MECHANISMS

One example of a compound reconfigurable aperture uses switches (based in solid-state, MEMS, or other technologies) to connect small subwavelength conductive pads that create any desired antenna topology [99]. The number of switches required for such an aperture can easily number into the thousands. A conceptual representation of the aperture is shown in Figure 7.1 [99]. In this case, connections between conductive pads are found using a genetic algorithm used in concert with a finite-difference time-domain full-wave electromagnetic simulator [99]. After each iteration of the genetic algorithm, the full-wave simulator is used to predict the performance of the resulting struc-

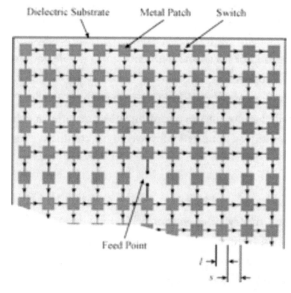

FIGURE 7.1: Conceptual drawing of a reconfigurable aperture antenna based on switched links between small metallic patches (Pringle et al. [99], ©IEEE 2004).

ture, and then the genetic algorithm makes more refinements to the design to achieve the specified performance goals. Although this flexible approach can deliver a range of structures with new and unexpected performance, it also has some limitations. In general, wideband configurations achieve less gain than narrowband configurations because wideband configurations implement fewer effective radiators in the aperture, which has a fixed physical size. Also, conductive pad density and switch capacitance may also degrade high-frequency operation [99]. Although the large number of switches between pads provides for graceful degradation of operation with switch failure, the bias network complexity for individually addressable switches may preclude practical use until the switches and bias networks can be directly integrated into the structure during fabrication.

Another version of a reconfigurable aperture uses the same conceptual approach as that in [99] but implements it in a different technology. In this case, semiconductor plasmas are used to form antenna structures [100]. High-conductivity plasma islands are formed and controlled by DC-injected currents into high-resistivity silicon-based diode structures. Figure 7.2 shows a detailed description of the proposed structure, including the plasma injection driver [100]. Of course, its implementation is just as complex as that of the previous example that uses many discrete switches. Although the creation of the necessary carrier densities in the silicon is a serious challenge, this approach is attractive for its ability to reduce its electromagnetic observability [100]. Essentially, when DC injection is stopped, the aperture "disappears." Because these approaches to compound reconfigurability face a large number of practical challenges for implementation on a large scale, they may be more appropriate to create intermediate apertures that are illuminated with more conventional sources, similar to the approach taken in [52]. Others have also proposed the use of controlled plasmas to achieve reconfigurable beam steering with leaky-wave structures [101].

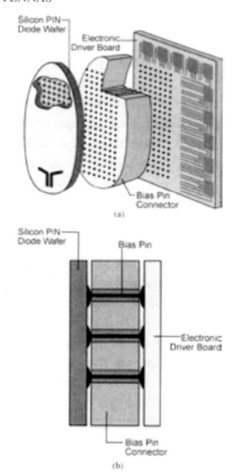

FIGURE 7.2: Depiction of a controllable plasma grid structure for a reconfigurable aperture. (a) The grid structure is fabricated on top of the silicon wafer. (b) A cross section of the plasma grid structure shows the layer interconnection (Fathy et al. [100], ©IEEE 2003).

•   •   •   •

CHAPTER 8

Practical Issues for Implementing Reconfigurable Antennas

8.1   TRADE-OFFS WITH RECONFIGURATION MECHANISM SELECTION

In many cases, the antenna topology and its intended application restrict the choice of reconfiguration mechanism. However, the designer still has some trade-offs to consider between available technologies. There are multiple dimensions in this trade-off space, including reconfiguration speed, power consumption, actuation requirements (voltage or current), fabrication complexity, durability, device lifetime, complexity of control and bias networks, weight, size, cost, dynamic range, sensitivity, and, of course, performance.

In general, mechanical reconfiguration mechanisms are slower than their electronic counterparts, but they can often achieve much more dramatic changes in performance through physical changes in the antenna structure. As electrically driven mechanical relays, RF-MEMS switches may promise better isolation and lower power consumption than solid-state switches such as PIN diodes and FETs, but they cannot yet compete in the dimensions of switching speed or power-handling capability. Other kinds of mechanical reconfiguration that require physical movement of the antenna components may provide more robust operation and be less complex to build, but they may also be sensitive to vibration or have very small tolerances to deliver the desired operating characteristics. All reconfiguration mechanisms will add weight, cost, and complexity compared with fixed antennas—the important trade-off dimension in this case will be the additional performance provided by reconfiguration that would be otherwise unattainable with a traditional design, which will be discussed more fully in Chapter 9.

8.2   RECONFIGURATION MECHANISM IMPLEMENTATION

Once a reconfiguration mechanism is chosen, its inclusion in the radiating structure poses one of the most challenging yet often ignored aspects of reconfigurable antenna development. Indeed, researchers often find it difficult to implement switching or control mechanisms into antennas that were originally developed using hard-wired connections for proof of concept. Certainly, careful and comprehensive design of the antenna and all of the necessary control hardware is required.

One excellent example of a reconfigurable antenna that includes the effects of the switching mechanism is provided by [20]. The basic antenna is a radiating slot that is reconfigured by changing its length with diodes across its width in several places. Transmission line-based models of the antenna that include the equivalent circuit model of the diodes help to predict the effects of the diodes and guide the adjustment of the antenna length and the placement of "matching switches." These matching switches compensate for any impedance mismatches caused by the nonideal effects of the switches used for frequency reconfiguration [20]. Switches based on RF-MEMS technology may also affect impedance and operating frequency, although not to as great as an extent as diodes or FETs, with the effects more pronounced at higher frequencies [10].

As another example, [82] provides an illustration of the complexities encountered when including prepackaged RF-MEMS switches in reconfigurable antennas. In this case, commercially available RF-MEMS switches with 50-Ω characteristic impedance were implemented on the surface of a radiation-reconfigurable spiral microstrip antenna. Because the microstrip line composing the antenna had a characteristic impedance much higher than 50 Ω, several significant changes to the original antenna design, the bias network design, and the switches themselves were necessary to achieve a working prototype [82]. Specifically, a tuning stub was added to the antenna structure, and the ground planes of the RF-MEMS switches were removed to decrease the effective impedance mismatch between the microstrip line and the switches [82]. As a result of the change to the switches, the required bias network for the switches also needed special attention. A diagram of the antenna and its associated bias networks for the switches are shown in Figures 8.1 and 8.2, respectively [82]. Other antenna designs have been designed specifically to accommodate switches with particular topologies and dimensions [e.g., 21].

One solution to this particular problem is to fabricate the antenna and any necessary switches in the same fabrication steps. Two examples of this approach can be found in [84] and [18]. In [84], a spiral microstrip design produced tilted radiation patterns with actuation of a subset of directly fabricated RF-MEMS switches that connect different portions of the spiral arms together. In [18], a multiband monopole antenna with integrated RF-MEMS switches delivered switched frequency bands. In both of these cases, details about switch and antenna packaging that address reliability and durability are still being resolved.

Often, adoption of new technologies to achieve the same kinds of reconfiguration can provide new levels of performance at reduced size, weight, and cost. One example of this is the evolution of the reconfigurable reflector antenna discussed in Chapter 6, developed first in [51, 52] and then later updated and improved in [53].

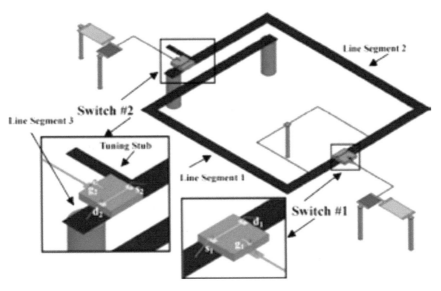

FIGURE 8.1: Pattern-reconfigurable microstrip antenna model using HFSS (TM) (Ansoft Corp., Pittsburgh, PA) including vias, lumped components, tuning and bias stubs, and simplified switch model including thin wires over silicon chips to approximate the switches. Antenna ground plane not shown (Huff and Bernhard [82], ©IEEE 2006).

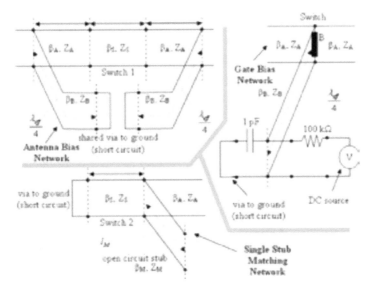

FIGURE 8.2: Equivalent transmission line circuits for determining the bias and matching networks for the antenna shown in Figure 8.1. The subscripts A, B, M, and S are used to reference propagation constants and impedances for the microstrip lines for the antenna, bias networks, matching network, and switch, respectively (Huff and Bernhard [82], ©IEEE 2006).

8.3   MATCHING NETWORKS

Extreme changes in operating frequency can lead to necessary changes in matching networks. Several authors (including [11]) specifically limit the reconfiguration range on frequency-reconfigurable antennas, for instance, so that a single-feed network can be used for all of the operating bands. A different approach is to make the impedance-matching network reconfigurable as well. Several research teams have developed novel and useful matching circuits that could be used in conjunction with reconfigurable antennas [e.g., 102–106]. Certainly, emerging applications such as cognitive radio can take advantage of the merging of these two developing technologies.

•   •   •   •

CHAPTER 9

Conclusions and Directions for Future Work

A review of the research discussed here indicates that reconfigurable antenna concepts and proto-types have been in development for more than 40 years. However, there are still very few reconfigurable antennas in use today. Why? There are several answers to this question, and they each point to a promising area of future research and development in this field.

9.1 QUANTIFICATION OF SYSTEM-LEVEL PERFORMANCE BENEFITS

First, the benefits and costs of implementation and operation of reconfigurable antennas are not clear from a system perspective. System and operating environment complexity makes it difficult to easily point to a particular antenna functionality that will automatically result in greater throughput, higher link reliability, or lower bit error rates. Certainly, as antenna engineers, we have instincts about this, but widespread recognition and adoption requires scientific, quantitative information that validates our instincts.

To date, there are only a handful of quantitative studies that consider the system-level performance benefits of antenna reconfigurability. One recent example uses pattern-reconfigurable dipole antennas to enable improvements in multiple-input multiple-output (MIMO) systems [107]. Here, a set of two printed frequency-reconfigurable dipoles spaced a quarter wavelength apart at 2.45 GHz were used at both the transmitter and receiver of a MIMO communication system. By changing the effective electrical lengths of the dipoles but keeping the operating frequency the same, a palette of four available patterns at each end of the communication link provided measurable capacity gains [107]. Packaging effects on the performance of pattern-reconfigurable antennas have also been investigated [108]. Another example of the demonstration of system-level benefits of antenna pattern reconfigurability is provided in [109]. Here, multiple pattern-reconfigurable antennas integrated into a laptop computer model provide increased spatial coverage over comparable fixed antennas for wireless local area network systems, resulting in the ability to adjust to changing electromagnetic environments and noise impairments [109].

Obviously, much more work needs to be done to form links between antenna functionality and system performance so that, eventually, particular reconfigurable antenna features will be specified to address known limitations or predicted operating scenarios. These links will also form the basis of new feedback hardware and processing that use information collected about the data, channel, or environment to select the best configuration for the antenna. This research also offers significant opportunities for collaboration with other engineers involved in the design of the necessary hardware, signal processing, and control algorithms that are necessary to take full advantage of antenna reconfigurability.

9.2   ECONOMIC VIABILITY OF RECONFIGURABILITY

Second, no clear cost/benefit methodology exists to assess the affordability and viability of reconfigurable antennas for new systems. Although reconfigurable antennas deliver expanded functionality over fixed antennas, they typically require extra parts, extra control lines, and extra infrastructure at one level or another. Antennas with all of these additional parts and connections are more expensive than fixed antennas. Without a clear cost/benefit analysis methodology in place, all of these "extras" may preclude inclusion of a reconfigurable antenna in a system simply because of budgetary constraints. In addition to making the case for a performance/cost trade-off, advances can be made in fabrication and system implementation that will make reconfigurable antennas more affordable in the future. One example of an approach that would result in more economical reconfigurable antennas is the direct integration of switches and associated control lines during antenna fabrication. Several research groups have started along this path, but significant fabrication and design challenges remain, including development of fabrication processes that can deliver high–aspect ratio structures as well as development of new reconfiguration mechanisms that are both compatible with antenna structures and capable of efficient and reliable operation.

9.3   EXPANDED DESIGN METHODOLOGIES

Third, other than in the most fundamental designs, there are few design guidelines or theories of operation for reconfigurable antennas that allow engineers to design their own antennas using the same principles. This means, in effect, that existing reconfigurable antennas are waiting for a suitable application to come along, rather than having applications dictate antenna functionality. Admittedly, the goal of being able to exactly specify and then deliver reconfigurability is an ambitious one. However, this goal is necessary to push the required technological and theoretical understanding of these structures to a point where many more reconfigurable antennas can be included in complex, multifunction systems. Certainly, with the expanding demands on wireless systems, we can look forward to new specifications for antenna functionality at multiple frequencies with varying bandwidth, polarization, and radiation characteristics.

9.4   NEW TECHNOLOGIES FOR RECONFIGURABILITY IN EMERGING APPLICATIONS

As new technologies are developed and mature, antenna engineers will work to include them in new designs. It is hoped that the needs of reconfigurable antennas can also spur development in several areas. Certainly, the maturation and validation of RF-MEMS technology is being pushed to a certain degree by demands for low-cost, high-reliability reconfigurable circuits and antennas. Additionally, pursuit of new and novel tunable materials with improved loss and bias characteristics will make reconfiguration more practical and cost-effective. Finally, the development of new kinds of mechanical actuators and electronic tuning methods can result in reconfigurable antennas with an even broader range of capabilities than those discussed here. With imagination, ingenuity, and enterprise, reconfigurable antennas can lead the way to new levels of wireless system performance.

• • • •

References

[1]   Bernhard, J.T., "Reconfigurable antennas and apertures: State-of-the-art and future outlook," *Proceedings of SPIE Conference on Smart Electronics, MEMS, BioMEMS, and Nanotechnology*, vol. 5055, pp. 1–9, 2003.

[2]   Brown, E.R., "On the gain of a reconfigurable-aperture antenna," *IEEE Transactions on Antennas and Propagation*, vol. 49, pp. 1357–1362, October 2001. doi:10.1109/8.954923

[3]   Pozar, D.M., *Microwave Engineering*, 2nd ed., John Wiley & Sons, New York, 1998.

[4]   Kraus, J.D., and Marhefka, R.J., *Antennas for All Applications*, 3rd ed., McGraw-Hill, New York, 2002.

[5]   *IEEE Standard Definition of Terms for Antennas*, IEEE Standard 145-1993, IEEE Press, New York, 1993.

[6]   Balanis, C.A., *Antenna Theory: Analysis and Design*, 3rd ed., Wiley Interscience, New York, 2005.

[7]   Filipovic, D.S., and Volakis, J.L., "A flush-mounted multifunctional slot aperture (combo-antenna) for automotive applications," *IEEE Transactions on Antennas and Propagation*, vol. 52, pp. 563–571, February 2004. doi:10.1109/TAP.2004.823927

[8]   Freeman, J.L., Lamberty, B.J., and Andrews, G.S., "Optoelectronically reconfigurable monopole antenna," *Electronics Letters*, vol. 28, no. 16, pp. 1502–1503, July 1992. doi:10.1049/el:19920954

[9]   Panagamuwa, C.J., Chauraya, A., and Vardaxoglou, J.C., "Frequency and beam reconfigurable antenna using photoconducting switches," *IEEE Transactions on Antennas and Propagation*, vol. 54, pp. 449–454, February 2006. doi:10.1109/TAP.2005.863393

[10]   Kiriazi, J., Ghali, H., Radaie, H., and Haddara, H., "Reconfigurable dual-band dipole antenna on silicon using series MEMS switches," *Proceedings of the IEEE/URSI International Symposium on Antennas and Propagation*, vol. 1, pp. 403–406, 2003.

[11]   Roscoe, D.J., Shafai, L., Ittipiboon, A., Cuhaci, M., and Douville, R., "Tunable dipole antennas," *Proceedings of the IEEE/URSI International Symposium on Antennas and Propagation*, vol. 2, pp. 672–675, 1993. doi:10.1109/APS.1993.385257

[12]   Weedon, W.H., Payne, W.J., and Rebeiz, G.M., "MEMS-switched reconfigurable antennas," *Proceedings of the IEEE/URSI International Symposium on Antennas and Propagation*, vol. 3, pp. 654–657, 2001.

[13]   Weedon, W., Payne, W., Rebeiz, G., Herd, J., and Champion, M., "MEMS-switched reconfigurable multi-band antenna: Design and modeling," *Proceedings of the 1999 Antenna Applications Symposium*, vol. 1, pp. 203–231, 1999.

[14] Ali, M.A., and Wahid, P., "A reconfigurable Yagi array for wireless applications," *Proceedings of the IEEE/URSI International Symposium on Antennas and Propagation*, vol. 1, pp. 466–468, 2002.

[15] Petko, J.S., and Werner, D.H., "Miniature reconfigurable three-dimensional fractal tree antennas," *IEEE Transactions on Antennas and Propagation*, vol. 52, pp. 1945–1956, August 2004. doi:10.1109/TAP.2004.832491

[16] Anagnostou, D., Chryssomallis, M.T., Lyke, J.C., and Christodoulou, C.G., "Re-configurable Sierpinski gasket antenna using RF-MEMS switches," *Proceedings of the IEEE/URSI International Symposium on Antennas and Propagation*, vol. 1, pp. 375–378, 2003.

[17] Vinoy, K., and Varadan, V., "Design of reconfigurable fractal antennas and RF-MEMS for space-based systems," *Smart Materials and Structures*, vol. 10, pp. 1211–1223, December 2001. doi:10.1088/0964-1726/10/6/310

[18] Anagnostou, D.E., Zheng, G., Chryssomallis, M.T., Lyke, J.C., Ponchak, G.E., Papapolymerou, J., and Christodoulou, C.G., "Design, fabrication, and measurements of an RF-MEMS–based self-similar reconfigurable antenna," *IEEE Transactions on Antennas and Propagation*, vol. 54, no. 2, pp. 422–432, February 2006.

[19] Gupta, K.C., Li, J., Ramadoss, R., and Wang, C., "Design of frequency-reconfigurable slot ring antennas," *Proceedings of the IEEE/URSI International Symposium on Antennas and Propagation*, vol. 1, p. 326, 2000.

[20] Peroulis, D., Sarabandi, K., and Katehi, L.P.B., "Design of reconfigurable slot antennas," *IEEE Transactions on Antennas and Propagation*, vol. 53, pp. 645–654, February 2005. doi:10.1109/TAP.2004.841339

[21] Huff, G.H., and Bernhard, J.T., "Frequency reconfigurable CPW-fed hybrid folded slot/slot dipole antenna," *Proceedings of the IEEE/ACES International Conference on Wireless Communications and Applied Computational Electromagnetics*, pp. 574–577, 2005.

[22] Yang, F., and Rahmat-Samii, Y., "Patch antenna with switchable slot (PASS): Dual frequency operation," *Microwave and Optical Technology Letters*, vol. 31, pp. 165–168, November 2001. doi:10.1002/mop.1388

[23] Wang, B.-Z., Xiao, S., and Wang, J., "Reconfigurable patch-antenna design for wideband wireless communication systems," *IET Microwaves, Antennas & Propagation*, vol. 1, no. 2, pp. 414–419, April 2007.

[24] Bhartia, P., and Bahl, I.J., "Frequency agile microstrip antennas," *Microwave Journal*, vol. 25, pp. 67–70, October 1982.

[25] Kawasaki, S., and Itoh, T., "A slot antenna with electronically tunable length," *Proceedings of the IEEE/URSI International Symposium on Antennas and Propagation*, vol. 1, pp. 130–133, 1991.

[26] Behdad, N., and Sarabandi, K., "A varactor-tuned dual-band slot antenna," *IEEE Transactions on Antennas and Propagation*, vol. 54, pp. 401–408, February 2006.

[27]   Behdad, N., and Sarabandi, K., "Dual-band reconfigurable antenna with a very wide tunability range," *IEEE Transactions on Antennas and Propagation*, vol. 54, pp. 409–416, February 2006. doi:10.1109/TAP.2005.863412

[28]   Erdil, E., Topalli, K., Unlu, M., Civi, O.A., and Akin, T., "Frequency tunable microstrip patch antenna using RF MEMS technology," *IEEE Transactions on Antennas and Propagation*, vol. 54, pp. 1193–1196, April 2007.

[29]   Shynu, S.V., Augustin, G., Aanandan, C.K., Mohanan, P., and Vasudevan, K., "C-shaped slot loaded reconfigurable microstrip antenna," *Electronics Letters*, vol. 42, no. 6, pp. 316–318, March 2006.

[30]   Jung, C.W., Kim, Y.J., Kim, Y.E., and De Flaviis, F., "Macro-micro frequency tuning antenna for reconfigurable wireless communication systems," *Electronics Letters*, vol. 43, pp. 201–202, February 15, 2007. doi:10.1049/el:20073906

[31]   Kiely, E., Washington, G., and Bernhard, J.T., "Design and development of smart microstrip patch antennas," *Smart Materials and Structures*, vol. 7, no. 6, pp. 792–800, December 1998. doi:10.1088/0964-1726/7/6/007

[32]   Kiely, E., Washington, G., and Bernhard, J.T., "Design, actuation, and control of active patch antennas," *Proceedings of the SPIE International Society for Optical Engineering*, vol. 3328, pp. 147–155, 1998.

[33]   Bernhard, J.T., Kiely, E., and Washington, G., "A smart mechanically-actuated two-layer electromagnetically coupled microstrip antenna with variable frequency, bandwidth, and antenna gain," *IEEE Transactions on Antennas and Propagation*, vol. 49, pp. 597–601, April 2001. doi:10.1109/8.923320

[34]   Langer, J.-C., Zou, J., Liu, C., and Bernhard, J.T., "Reconfigurable out-of-plane microstrip patch antenna using MEMS plastic deformation magnetic actuation," *IEEE Microwave and Wireless Components Letters*, vol. 13, pp. 120–122, March 2003.

[35]   Pozar, D.M., and Sanchez, V., "Magnetic tuning of a microstrip antenna on a ferrite substrate," *Electronics Letters*, vol. 24, pp. 729–731, June 9, 1988. doi:10.1049/el:19880491

[36]   Mishra, R.K., Pattnaik, S.S., and Das, N., "Tuning of microstrip antenna on ferrite substrate," *IEEE Transactions on Antennas and Propagation*, vol. 41, pp. 230–233, February 1993. doi:10.1109/8.214616

[37]   Brown, A.D., Volakis, J.L., Kempel, L.C., and Botros, Y., "Patch antennas on ferromagnetic substrates," *IEEE Transactions on Antennas and Propagation*, vol. 47, pp. 26–32, January 1999. doi:10.1109/8.752980

[38]   Romanofsky, R.R., Miranda, F.A., Van Keuls, F.W., and Valerio, M.D., "Recent advances in microwave applications of thin ferroelectric films at the NASA Glenn Research Center," *Materials Research Society Symposium Proceedings*, vol. 833, pp. 173–181, 2004.

[39]   Miranda, F.A., Van Keuls, F.W., Romanofsky, R.R., Mueller, C.H., Alterovitz, S., and Subramanyam, G., "Ferroelectric thin films-based technology for frequency- and phase-agile

microwave communication applications," *Integrated Ferroelectrics*, vol. 42, pp. 131–149, 2002. doi:10.1080/10584580210845

[40] Xu, H., Pervez, N.K., and York, R.A., "Tunable microwave integrated circuits BST thin film capacitors with device structure optimization," *Integrated Ferroelectrics*, vol. 77, pp. 27–35, 2005. doi:10.1080/10584580500413681

[41] Boti, M., Dussopt, L., and Laheurte, J.-M., "Circularly polarized antenna with switchable polarization sense," *Electronics Letters*, vol. 36, pp. 1518–1519, August 2000. doi:10.1049/el:20001098

[42] Yang, F., and Rahmat-Samii, Y., "Patch antennas with switchable slots (PASS) in wireless communications: Concepts, designs, and applications," *IEEE Antennas and Propagation Magazine*, vol. 47, pp. 13–29, April 2005. doi:10.1109/MAP.2005.1487774

[43] Yang, F., and Rahmat-Samii, Y., "A reconfigurable patch antenna using switchable slots for circular polarization diversity," *IEEE Microwave and Wireless Components Letters*, vol. 12, pp. 96–98, March 2002. doi:10.1109/7260.989863

[44] Sung, Y.J., Jang, T.U., and Kim, Y.-S., "A reconfigurable microstrip antenna for switchable polarization," *IEEE Microwave and Wireless Components Letters*, vol. 14, pp. 534–536, November 2004. doi:10.1109/LMWC.2004.837061

[45] Fries, M.K., Grani, M., and Vahldieck, R., "A reconfigurable slot antenna with switchable polarization," *IEEE Microwave and Wireless Components Letters*, vol. 13, pp. 490–492, November 2003. doi:10.1109/LMWC.2003.817148

[46] Simons, R.N., Chun, D., and Katehi, L.P.B., "Polarization reconfigurable patch antenna using microelectromechanical systems (MEMS) actuators," *Proceedings of the IEEE/URSI International Symposium on Antennas and Propagation*, vol. 1, pp. 6–9, 2002.

[47] Hsu, S.-H., and Chang, K., "Novel reconfigurable microstrip antenna with switchable circular polarization," *IEEE Antennas and Wireless Propagation Letters*, vol. 6, pp. 160–162, 2007. doi:10.1109/LAWP.2007.894150

[48] Rainville, P., and Harackiewicz, F., "Magnetic tuning of a microstrip patch antenna fabricated on a ferrite film," *IEEE Microwave and Wireless Components Letters*, vol. 2, pp. 483–485, December 1992.

[49] Yang, S.-L.S., and Luk, K.-M., "Design of a wide-band L-probe patch antenna for pattern reconfiguration or diversity applications," *IEEE Transactions on Antennas and Propagation*, vol. 54, pp. 433–438, February 2006. doi:10.1109/TAP.2005.863376

[50] Clarricoats, P.J.B., and Zhou, H., "The design and performance of a reconfigurable mesh reflector antenna," *IEE Seventh International Conference on Antennas and Propagation*, vol. 1, pp. 322–325, 1991.

[51] Clarricoats, P.J.B., Zhou, H., and Monk, A., "Electronically controlled reconfigurable reflector antenna," *Proceedings of the IEEE/URSI International Symposium on Antennas and Propagation*, vol. 1, pp. 179–181, 1991.

[52] Washington, G., Yoon, H.-S., Angelino, M., and Theunissen, W.H., "Design, modeling, and optimization of mechanically reconfigurable aperture antennas," *IEEE Transactions on Antennas and Propagation*, vol. 50, pp. 628–637, May 2002.

[53] Sievenpiper, D., Schaffner, J., Loo, R., Tangonan, G., Ontiveros, S., and Harold, R. "A tunable impedance surface performing as a reconfigurable beam steering reflector," *IEEE Transactions on Antennas and Propagation*, vol. 50, pp. 384–390, March 2002.

[54] Sievenpiper, D.F., Schaffner, J.H., Song, H.J., Loo, R.Y., and Tangonan, G., "Two-dimensional beam steering using an electrically tunable impedance surface," *IEEE Transactions on Antennas and Propagation*, vol. 51, pp. 2713–2722, October 2003.

[55] Sievenpiper, D., Schaffner, J., Lee, J.J., and Livingston, S., "A steerable leaky-wave antenna using a tunable impedance ground plane," *IEEE Antennas and Wireless Propagation Letters*, vol. 1, pp. 179–182, 2002. doi:10.1109/LAWP.2002.807788

[56] Sievenpiper, D., and Schaffner, J., "Beam steering microwave reflector based on electrically tunable impedance surface," *Electronics Letters*, vol. 38, no. 21, pp. 1237–1238, October 10, 2002. doi:10.1049/el:20020863

[57] Sievenpiper, D.F., "Forward and backward leaky wave radiation with large effective aperture from an electronically tunable textured surface," *IEEE Transactions on Antennas and Propagation*, vol. 53, pp. 236–247, January 2005. doi:10.1109/TAP.2004.840516

[58] Rotman, W., and Maestri, A., "An electromechanically scannable trough waveguide array," *Proceedings of the IRE International Convention Record*, vol. 8, pp. 67–83, March 1960. doi:10.1109/IRECON.1960.1150886

[59] Rotman, W., and Oliner, A., "Asymmetrical trough waveguide antennas," *IEEE Transactions on Antennas and Propagation*, vol. 7, pp. 153–162, April 1959.

[60] Huff, G.H., and Bernhard, J.T., "Electromechanical beam steering of a trough waveguide antenna using cantilever perturbations," *Proceedings of the 2005 Antenna Applications Symposium*, pp. 152–165, 2005.

[61] Chiao, J.-C., Yiton, F., Iao, M.C., DeLisio, M., and Lin, L.-Y., "MEMS reconfigurable Vee antenna," *IEEE MTT-S International Microwave Symposium Digest*, vol. 4, pp. 1515–1518, 1999.

[62] Chang, K., Li, M., Yun, T.-Y., and Rodenbeck, C.T., "Novel low-cost beam steering techniques," *IEEE Transactions on Antennas and Propagation*, vol. 50, pp. 618–627, May 2002. doi:10.1109/TAP.2002.1011227

[63] Rodenbeck, C.T., Li, M., and Chang, K., "Design and analysis of a reconfigurable dual-beam grating antenna for low-cost millimeter-wave beam-steering," *IEEE Transactions on Antennas and Propagation*, vol. 52, pp. 999–1006, April 2004. doi:10.1109/TAP.2004.825676

[64] Nikolaou, S., Bairavasubramanian, R., Lugo, Jr., C., Carrasquillo, I., Thompson, D.C., Ponchak, G.E., Papapolymerou, J., and Tentzeris, M.M., "Pattern and frequency reconfigurable annular slot antenna using PIN diodes," *IEEE Transactions on Antennas and Propagation*, vol. 54, pp. 439–448, February 2006. doi:10.1109/TAP.2005.863398

[65] Harrington, R.F., "Reactively controlled directive arrays," *IEEE Transactions on Antennas and Propagation*, vol. 26, pp. 390–395, May 1978. doi:10.1109/TAP.1978.1141852

[66] Luzwick, J., and Harrington, R., "A reactively loaded aperture antenna array," *IEEE Transactions on Antennas and Propagation*, vol. 26, no. 4, pp. 543–547, July 1978.

[67] Preston, S.L., Thiel, D.V., Lu, J.W., O'Keefe, S.G., and Bird, T.S., "Electronic beam steering using switched parasitic patch elements," *Electronics Letters*, vol. 33, no. 1, pp. 7–8, January 1997. doi:10.1049/el:19970048

[68] Schlub, R., Thiel, D.V., Lu, J.W., and O'Keefe, S.G., "Dual-band six-element switched parasitic array for smart antenna cellular communications," *Electronics Letters*, vol. 36, no. 16, pp. 1342–1343, August 2000. doi:10.1049/el:20000995

[69] Thiel, D.V., and Smith, S., *Switched Parasitic Antennas for Cellular Communications*, Artech House, Boston, MA, 2002.

[70] Thiel, D.V., "Switched parasitic antennas and controlled reactance parasitic antennas: A systems comparison," *Proceedings of the IEEE/URSI Antennas and Propagation International Symposium*, vol. 3, pp. 3211–3214, 2004.

[71] Thiel, D.V., "Impedance variations in controlled reactance parasitic antennas," *Proceedings of the IEEE International Symposium on Antennas and Propagation*, vol. 3A, pp. 671–674, 2005.

[72] Ohira, T., and Gyoda, K., "Hand-held microwave direction-of-arrival finder based on varactor-tuned analog aerial beamforming," *Proceedings of the IEEE Asia Pacific Conference* (Taipei), vol. 2, pp. 585–588, 2001.

[73] Schlub, R., Lu, J., and Ohira, T., "Seven-element ground skirt monopole ESPAR antenna design from a genetic algorithm and finite element method," *IEEE Transactions on Antennas and Propagation*, vol. 51, no. 11, pp. 3033–3039, November 2003. doi:10.1109/TAP.2003.818790

[74] Ohira, T., and Iigusa, K., "Electronically steerable array radiator antenna," Transactions of the IEICE, vol. J87-C, pp. 12–31, 2004.

[75] Zhang, S., Huff, G.H., Feng, J., and Bernhard, J.T., "A pattern reconfigurable microstrip parasitic array," *IEEE Transactions on Antennas and Propagation*, vol. 52, pp. 2773–2776, October 2004. doi:10.1109/TAP.2004.834372

[76] Zhang, S., Huff, G., Cung, G., and Bernhard, J.T., "Three variations of a pattern reconfigurable microstrip parasitic array," *Microwave and Optical Technology Letters*, vol. 45, pp. 369–372, June 2005. doi:10.1002/mop.20826

[77] Zhang, S., *A Pattern Reconfigurable Microstrip Parasitic Array: Theory, Design, and Applications*, PhD dissertation, University of Illinois at Urbana–Champaign, 2005.

[78] Dinger, R.J., "Reactively steered adaptive array using microstrip patch elements at 4 GHz," *IEEE Transactions on Antennas and Propagation*, vol. 32, pp. 848–856, August 1984. doi:10.1109/TAP.1984.1143420

[79] Dinger, R.J., "A planar version of a 4.0 GHz reactively steered adaptive array," *IEEE Transactions on Antennas and Propagation*, vol. 34, pp. 427–431, March 1986.

[80] Huff, G.H., Feng, J., Zhang, S., and Bernhard, J.T., "A novel radiation pattern and frequency reconfigurable single turn square spiral microstrip antenna," *IEEE Microwave and Wireless Components Letters*, vol. 13, pp. 57–59, February 2003. doi:10.1109/LMWC.2003.808714

[81] Huff, G.H., and Bernhard, J.T., "Analysis of a radiation and frequency reconfigurable microstrip antenna," *Proceedings of the 2004 Antenna Applications Symposium*, pp. 175–191, September 2004.

[82] Huff, G.H., and Bernhard, J.T., "Integration of packaged RF MEMS switches with radiation pattern reconfigurable square spiral microstrip antennas," *IEEE Transactions on Antennas and Propagation*, vol. 54, pp. 464–469, February 2006. doi:10.1109/TAP.2005.863409

[83] Chen, S.-H., Row, J.-S., and Wong, K.-L., "Reconfigurable square-ring patch antenna with pattern diversity," *IEEE Transactions on Antennas and Propagation*, vol. 55, pp. 472–475, February 2007.

[84] Jung, C., Lee, M., Li, G.P., and De Flaviis, F., "Reconfigurable scan-beam single-arm spiral antenna integrated with RF-MEMS switches," *IEEE Transactions on Antennas and Propagation*, vol. 54, pp. 455–463, February 2006. doi:10.1109/TAP.2005.863407

[85] Mehta, A., Mirshekar-Syahkal, D., and Nakano, H., "Beam adaptive single arm rectangular spiral antenna with switches," *IEE Proceedings—Microwaves, Antennas and Propagation*, vol. 153, no. 1, pp. 13–18, February 2006. doi:10.1049/ip-map:20050045

[86] Riel, M., and Laurin, J.-J., "Design of an electronically beam scanning reflectarray using aperture-coupled elements," *IEEE Transactions on Antennas and Propagation*, vol. 55, pp. 1260–1266, May 2007. doi:10.1109/TAP.2007.895586

[87] Henderson, A., James, J.R., Fray, A., and Evans, G.D., "New ideas for beam scanning using magnetised ferrite," *Proceedings of the IEE Colloquium on Electronically Scanned Antennas*, vol. 1, January 21, 1988.

[88] Lovat, G., Burghignoli, P., and Celozzi, S., "A tunable ferroelectric antenna for fixed-frequency scanning applications," *IEEE Antennas and Wireless Propagation Letters*, vol. 5, pp. 353–356, 2006.

[89] Yashchyshyn, Y., and Modelski, J.W., "Rigorous analysis and investigations of the scan antennas on a ferroelectric substrate," *IEEE Transactions on Microwave Theory and Techniques*, vol. 53, pp. 427–438, February 2005. doi:10.1109/TMTT.2004.840779

[90] Varadan, V.K., Varadan, V.V., Jose, K.A., and Kelly, J.F., "Electronically steerable leaky wave antenna using a tunable ferroelectric material," *Smart Materials and Structures*, vol. 3, pp. 470–475, 1994. doi:10.1088/0964-1726/3/4/009

[91] Iskander, M.F., Zhang, Z., Yun, Z., Jensen, R., and Redd, S., "Design of a low-cost 2-D beam-steering antenna using ferroelectric material and CTS technology," *IEEE Transactions on Microwave Theory and Techniques*, vol. 49, pp. 1000–1003, May 2001.

[92] Borowick, J., and Stern, R., "A line source array for limited scan applications," *Proceedings of the IEEE/URSI International Symposium on Antennas and Propagation*, vol. 1, pp. 164–167, October 1976.

[93] Maheri, H., Tsutsumi, M., and Kumagai, N., "Experimental studies of magnetically scannable leaky wave antennas having corrugated ferrite slab/dielectric layer structure," *IEEE Transactions on Antennas and Propagation*, vol. 36, pp. 911–917, July 1988.

[94] Sun, L., How, H., Vittoria, C., and Champion, M., "Steerable linear dipole antenna using ferrite phase shifters," *IEEE Transactions on Magnetics*, vol. 33, pp. 3436–3438, December 1997. doi:10.1109/20.617969

[95] Batchelor, J., and Langley, R., "Beam scanning using microstrip line on biased ferrite," *Electronics Letters*, vol. 33, no. 8, pp. 645–646, April 1997.

[96] Brown, A.D., Kempel, L.C., and Volakis, J.L., "Design method for antenna arrays employing ferrite printed transmission line phase shifters," *IEE Proceedings—Microwaves, Antennas and Propagation*, vol. 149, no. 1, pp. 33–40, February 2002. doi:10.1049/ip-map:20020146

[97] Yang, X.-S., Wang, B.-Z., Wu, W., and Xiao, S., "Yagi patch antenna with dual-band and pattern reconfigurable characteristics," *IEEE Antennas and Wireless Propagation Letters*, vol. 6, pp. 168–171, 2007. doi:10.1109/LAWP.2007.895292

[98] Ali, M., Sayem, A.T.M., and Kunda, V.K., "A reconfigurable stacked microstrip patch antenna for satellite and terrestrial links," *IEEE Transactions on Vehicular Technology*, vol. 56, pp. 426–435, March 2007.

[99] Pringle, L.N., Harms, P.H., Blalock, S.P., Kiesel, G.N., Kuster, E.J., Friederich, P.G., Prado, R.J., Morris, J.M., and Smith, G.S., "A reconfigurable aperture antenna based on switched links between electrically small metallic patches," *IEEE Transactions on Antennas and Propagation*, vol. 52, pp. 1434–1445, June 2004.

[100] Fathy, A., Rosen, A., Owen, H., McGinty, F., McGee, D., Taylor, G., Amantea, R., Swain, P., Perlow, S., and ElSherbiny, M., "Silicon-based reconfigurable antennas—Concepts, analysis, implementation, and feasibility," *IEEE Transactions on Microwave Theory and Techniques*, vol. 51, pp. 1650–1661, June 2003. doi:10.1109/TMTT.2003.812559

[101] Grewal, G., and Hanson, G.W., "Optically-controlled solid-state plasma leaky-wave antenna," *Microwave and Optical Technology Letters*, vol. 39, pp. 450–453, December 2003. doi:10.1002/mop.11245

[102] Papapolymerou, J., Lange, K.L., Goldsmith, C.L., Malczewski, A., and Kleber, J., "Reconfigurable double-stub tuners using MEMS switches for intelligent RF front-ends," *IEEE Transactions on Microwave Theory and Techniques*, vol. 51, pp. 271–278, January 2003.

[103] Lu, Y., Peroulis, D., Mohammadi, S., and Katehi, L.P.B., "A MEMS reconfigurable matching network for a class AB amplifier," *IEEE Microwave and Wireless Components Letters*, vol. 13, pp. 437–439, October 2003. doi:10.1109/LMWC.2003.818523

[104] Sjoblom, P., and Sjoland, H., "An adaptive impedance tuning CMOS circuit for ISM 2.4-GHz band," *IEEE Transactions on Circuits and Systems* I, vol. 52, pp. 1115–1124, June 2005. doi:10.1109/TCSI.2005.849116

[105] Whatley, R.B., Zhou, Z., and Melde, K.L., "Reconfigurable RF impedance tuner for match control in broadband wireless devices," *IEEE Transactions on Antennas and Propagation*, vol. 54, pp. 470–478, February 2006. doi:10.1109/TAP.2005.863396

[106] Zhou, Z., and Melde, K.L., "Frequency agility of broadband antennas integrated with a reconfigurable RF impedance tuner," *IEEE Antennas and Wireless Propagation Letters*, vol. 6, pp. 56–59, 2007. doi:10.1109/LAWP.2007.891955

[107] Piazza, D., and Dandekar, K.R., "Reconfigurable antenna solution for MIMO-OFDM systems," *Electronics Letters*, vol. 42, no. 8, pp. 446–447, April 2006. doi:10.1049/el:20060221

[108] Huff, G.H., Feng, J., Zhang, S., and Bernhard, J.T., "Directional reconfigurable antennas on laptop computers: Simulation, measurement, and evaluation of candidate integration positions." *IEEE Transactions on Antennas and Propagation*, vol. 52, pp. 3220–3227, December 2004.

[109] Roach, T.L., Huff, G.H., and Bernhard, J.T., "A comparative study of diversity gain and spatial coverage: Fixed versus reconfigurable antennas for portable devices," *Microwave and Optical Technology Letters*, vol. 49, no. 3, pp. 535–539, March 2007. doi:10.1002/mop.22181

Author Biography

Jennifer T. Bernhard was born on May 1, 1966, in New Hartford, NY. She received her bachelor of science degree in electrical engineering from Cornell University in 1988. She received her master of science and doctor of philosophy degrees in electrical engineering from Duke University in 1990 and 1994, respectively, with support from a National Science Foundation Graduate Fellowship.

During the 1994–1995 academic year, she held the position of postdoctoral research associate with the Departments of Radiation Oncology and Electrical Engineering at Duke University, where she developed circuitry for simultaneous hyperthermia and magnetic resonance imaging thermometry. At Duke, she was also an organizing member of the Women in Science and Engineering (WISE) Project, a graduate student-run organization designed to improve the climate for graduate women in engineering and the sciences. From 1995 to 1999, Prof. Bernhard was an assistant professor in the Department of Electrical and Computer Engineering at the University of New Hampshire, where she held the Class of 1944 Professorship. From 1999 to 2003, she was an assistant professor in the Department of Electrical and Computer Engineering at University of Illinois at Urbana–Champaign. Since 2003, she has held the position of associate professor at Illinois.

In 1999 and 2000, she was a NASA-ASEE Summer Faculty Fellow at NASA Glenn Research Center in Cleveland, OH. Prof. Bernhard received the NSF CAREER Award in 2000. She and her students received the 2004 H. A. Wheeler Applications Prize Paper Award from the IEEE Antenna and Propagation Society for their article published in the March 2003 issue of the *IEEE Transactions on Antennas and Propagation*. Prof. Bernhard's research interests include reconfigurable and wideband microwave antennas and circuits, wireless sensors and sensor networks, high-speed wireless data communication, electromagnetic compatibility, and electromagnetics for industrial, agricultural, and medical applications. She served as an associate editor for *IEEE Transactions on Antennas and Propagation* from 2001 to 2007 and as an associate editor for *IEEE Antennas and Wireless Propagation Letters* from 2001 to 2005. She is also a member of the editorial board of *Smart Structures and Systems*. She is a member of URSI Commissions B and D, Tau Beta Pi, Eta Kappa Nu, Sigma Xi, and ASEE. She is a senior member of the IEEE and served as an elected member of the IEEE Antennas and Propagation Society's Administrative Committee from 2004 to 2006. She is 2007 president-elect of the IEEE Antennas and Propagation Society.

Printed in the United States
by Baker & Taylor Publisher Services